Excellence —
The IBM Way

Excellence — The IBM Way

H. James Harrington

ASQC Quality Press
American Society for Quality Control
310 West Wisconsin Avenue
Milwaukee, Wisconsin 53203

Published by ASQC Quality Press

Milwaukee

Excellence — The IBM Way

H. James Harrington

ISBN 0-87389-037-X

Foreword

It comes as a shock when a company that takes pride in its service realizes that a significant competitor is out to show, and with justification, that "service" is a bad word when it can be replaced with "no need for service." It comes as a shock when a company that prides itself on stringent quality targets realizes that its competitors are doing better, that some of its own manufacturing facilities can do dramatically better than others, and that the targets themselves become security blankets shielding the target owners from the constant need to analyze and improve. It comes as a shock for a company that believes it has a great understanding of its cost structures when that company begins to do a thoughtful job of measuring the "cost of quality" (or lack thereof) to find that major elements of cost have been unrealized and poorly controlled. It comes as a shock when a company thought process, which has equated "quality" with what happens only in manufacturing, is shown to be a deterent to dramatic improvement in all sectors of the company business.

These "shocks," and many others of like nature, became the major stimulus to management action in the late 1970s as Jim Harrington's book begins. First, the management became firmly convinced of the need to improve across the board. And second, serious evaluations — taken in an affirmative desire for methods of improvement — quickly showed numerous impediments and opportunities. The programs Jim describes were put in place with dramatic results.

When we started, quality was a manufacturing issue. As we proceeded it became a company issue — management, design, development, administration, programming, service, marketing, manufacturing . . . . Everyone makes a "product" and the processes of making those products can be understood and improved. When we started, areas such as programming and hardware design openly thought we were daft for talking about "zero defects" as a target. When we started, most functions argued that the quality of "their products" was not measurable.

Now, after the major part of a decade, much learning, and thoughtful leadership from individuals like Jim Harrington, it's a different world. I know that IBM will never go back to the relaxed, expensive, limited view of quality.

Arthur G. Anderson
Former President
General Products Division
IBM Corporation

This book is dedicated to the man who has always set a high standard of excellence and has dedicated his life to IBM. There is no better example of a good man and a good IBMer — my father, Frank O. Harrington.

Contents

x

Abstract

This book is an overview of the steps the IBM Corporation has taken since 1980 to improve the excellence of each employee's output and of the final product. Individual operating units customized the improvement process to meet their own needs. As a result, many of the operating units implemented only some of the activities presented here, while others did more. This book also includes a number of case studies that demonstrate how effective the improvement process has been in both white-collar and blue-collar areas.

Part One
A Plan for Excellence

Introduction

Although IBM has long been a leader in product and service quality, setting the world quality standard for computer operations, we realize that any company cannot sustain its leadership for long in any area unless it continues to improve. The methods and techniques that made a company great in the 1960s just met the requirements of the 1970s, and are inadequate in the 1980s.

As we looked back at the 1970s, we believed we had been successful in sustaining our quality reputation. But as Samuel Rutgers said back in the 17th century, "It takes real talent to withstand adversity, but it takes genius to survive success." At IBM we not only wanted to survive success, but we also wanted to prosper in the face of it. To accomplish this we looked at quality as one of our strongest business assets. We also realized that continuing improvement was a necessity if we were to survive, because competition in the United States and overseas was increasing and improving rapidly. After looking closely at the competition and IBM's internal operations, we defined quality improvement as one of our best opportunities to reduce costs and increase customer satisfaction. John F. Akers, president of IBM, summed it all up in five words: "Quality is the competitive edge."

For years IBM had been known for its outstanding service. When a customer had a problem IBM would darken the skies with planes filled with people rushing to help that customer. It was a strategy that worked. As a result the service portion of the corporation was among the fastest growing parts of the business. Although we did not want to reduce the service we provided to our customers, we wanted to minimize their need for it, and thereby to become known not for the service we provided, but for not needing to provide it. Error-free products require no service.

We also believed that improved quality provided the best opportunity to reduce costs, increase profits, and provide less expensive products to our customers. Although our performance records were and are unmatched throughout the computer industry, it was costing us far too much to provide this superior performance. In many cases, the cost of poor quality accounted for 40 percent of the product cost. By reducing our cost of poor quality, we knew we could increase profits, reduce product costs, and increase product performance, thereby benefiting our customers, our stockholders, and our employees.

Since 1980, our progress in improving quality has been rapid, and our returns on investment have far surpassed our original estimates; but we have not reached the end of our journey. In fact, we are just at the beginning of it, and we have a long way to go before we reach our full potential. Certainly, we cannot relax our focus on improvement, for the moment we relax — the moment we redirect our attention away from continuous improvement — is the moment we will start to slip backward.

Has the improvement process worked for IBM? We believe it has because total costs have been reduced significantly, profits have increased, customer satisfaction has reached an all-time high, and employees are more satisfied with the work they are doing. Most important of all, IBM is still viewed as the world leader in the data processing field, and we set the world-class standard for excellence in this field.

The worldwide acceptance of IBM excellence has been verified at least twice in the last 12 months. First, when nearly 700 senior executives were polled by the Gallup Corporation and asked what companies they associated with high quality, they named more than 200 different American, Asian, and European firms. But one company was named more than twice as often as any other — IBM. Second, in 1987 when NASA first selected winners for their NASA Excellence Award for the application of quality methods and quality improvement in all areas of the business, IBM was one of the two companies selected to receive the award. Based on these indicators of corporate excellence, I hope this IBM case study will be useful to help companies around the world provide higher quality products and services at lower prices to their customers.

The culture of a corporation is based on its history, its beliefs, its traditions, its successes, and even its failures. Living within this culture, leaders impose their individual personalities on the corporate personality, which causes superficial changes to occur for short periods of time. At IBM we have a corporate culture we are very proud of, one that has excellence and service to our customers as two of its corporate cornerstones. This culture has made excellence a way of life at IBM. It is part of everything we do, every decision we make. It is a long-standing tradition. Thomas J. Watson, Jr., stated, "We believe that all organizations should pursue all tasks with the idea that they can be accomplished in a superior fashion."

The IBM culture is one of helping and expecting each employee to do the very best each minute of each day; for it is only in helping the employee to grow and develop that the corporation will prosper. The corporation with a people-building culture will prosper, and the corporation with a people-using culture will fail in the long run. In the book entitled *A Business and its Beliefs*, Thomas J. Watson, Jr., wrote, "Our emphasis on human relations was not motivated by altruism, but by the simple belief that if we respect our people and help them to respect themselves, the company will make the most profit."

Excellence does not happen by itself. It is not like tomorrow; you cannot just stand around and wait for it to come. Excellence requires a well-thought-out plan. I like to think of the planning process as a pyramid made up of six levels (Figure 1).

Figure 1. The Planning Pyramid

1. *Mission.* This is the reason the corporation exists. It is developed by the president of the company and is the reason stockholders invest in the company. The mission statement for IBM could be simply worded as: "To service the worldwide information handling needs."

2. *Operating principles (basic beliefs).* These are the fundamental beliefs on which corporate culture is based. They keep good employees and attract new employees. The operating principles are developed by top corporate management and change infrequently. IBM's basic beliefs are:

  • Respect for the individual's rights and dignity. This means that IBM will help its employees develop their full potential and make the best use of their abilities. It also indicates that IBM will pay and promote individuals based on merit, not on seniority.
  • Service to our customers. Our products and service bring profits only to the degree that they serve the customer and satisfy customer needs.

- Excellence must be a way of life. We want IBM to be known for its excellence. Therefore, we believe that every task, in every part of the business, should be performed in a superior manner and to the best of our ability. Nothing should be left to chance in our pursuit of excellence.
- Managers must lead effectively. Our success depends on intelligent and aggressive managers who are sensitive to the need for making an enthusiastic partner of every individual in the organization.
- Obligation to the stockholders. IBM has an obligation to its stockholders to take care of the property they have entrusted to us, and to provide an attractive return on invested capital.
- Fair deal for the business associate (supplier). IBM selects business associates (suppliers) in accordance with the quality of their products and services, their general reliability, and competitiveness of price. (Note: To recognize the important interdependencies among IBM and the companies that supply us with components and services, IBM frequently refers to suppliers as business associates.)
- IBM should be a good corporate citizen. This belief recognizes the responsibilities that a corporation has to the city, state, and country where it is located and with which it does business.

These basic beliefs, used to guide both our internal and external activities, show that IBM has made excellence an integral part of its corporate culture.

3. *Business objectives.* Business objectives set the corporation's direction over a period of time. They are a well-publicized set of objectives that provide management and employees with information related to what the corporation wants to accomplish in the next 10 to 20 years. In the latter part of the 1970s, IBM released its objectives for the 1980s. In 1987 one additional goal, directed at service to our customers, was added. These goals are listed as follows:

- To enhance our customer partnerships.
- To grow with the industry.
- To exhibit product leadership across our entire product line. To excel in technology, value, and quality.
- To be the most effective in everything we do. To be the low-cost producer, the low-cost seller, the low-cost administrator.
- To sustain our profitability, which funds our growth.

After studying the situation and many alternatives, we realized that one of the best ways to accomplish these objectives was by improving product and service quality levels. As John Akers said, "From IBM's standpoint, quality improvement is at the very heart of our business goals for the 1980s. We are convinced that quality is central to the achievement of these goals."

Top management develops and publishes business objectives. They are subject to change as the business climate changes, and as specific objectives are met.

4. *Performance goals.* These can take the form of short- and long-range targets that support the business objectives. They should be quantifiable, measurable, and time related. (Example: Increase sales at a minimum rate of X percent per year from 1988 to 1998, with an average annual growth rate of Y percent.) A typical long-range performance goal would be to decrease the cost of maintaining customer-purchased equipment at a minimum rate of 10 percent per year for the next five years; or to correct 99.7 percent of all customer problems with one service call over the next 24 months. Each year a set of short-range goals should be generated by first-line and middle management in each function. These commitments should be directly tied into their budgets. These goals should be reviewed and approved by top management to ensure that they are aggressive enough and support business objectives. (Note: Goals have two key ingredients. First, they specifically state the target for improvement; and second, they give the time interval in which the improvement will be accomplished.)

5. *Strategy.* The strategy documents the approach that will be used to meet business goals. It is generated by middle management and approved by upper management. Every effort should be made to keep the strategy up-to-date, without making major changes. Drastic changes in the strategy upset the corporation and require a major expenditure of resources in reacting to those changes. Major changes can also result in the termination of projects that are only partially completed and/or have not become totally effective. It should be apparent that many strategies are generated by many different functions, supporting the business objectives. For the purposes of this book, we will concentrate on only the quality strategies.

Many strategies were considered by IBM: burn-in programs, process qualification systems, expanded education, etc. Each had its own advocates, and each was good in its own right. In the long run the strategy that best fit our total needs was a simple one: the integration of quality responsibilities into the activities of each area. This strategy had the advantage of not increasing the size of our quality assurance organizations and of aligning accountability with responsibility.

6. *Tactics.* Tactics are the "how to's." They are the actions planned or being taken to meet performance goals. Tactics are generated by first-line and middle management and are approved by upper management. They are updated at least once a year and change frequently, based on experience and business needs. Employees in the first-line departments are encouraged to participate in the preparation of tactics, as they will eventually be responsible for implementing them.

Note that the mission, operating principles, and business objectives come down from the top. Strategy, tactics, and performance objectives come up from the bottom, meeting in a common understanding of the long-range plan and a total commitment to it. It is absolutely imperative that the long-range plan be a living, breathing thing understood and supported by every member of the organization.

IBM has had a very effective quality assurance organization at least since the 1920s. This organization was among the leaders in developing many of the present quality assurance tools, i.e., process qualification, stress testing, customer reporting, indirect poor-quality costs, etc.

This was and is a very effective organization with all the ingredients of a total quality assurance organization. It did failure analysis, early entry, manufacturability studies, statistical process control, customer interface analyses, just to mention a few activities. Its engineers and inspectors evaluated, measured, controlled, and reported on the total process cycle. This, along with the efforts of the total IBM team, made IBM the standard for excellence in the computer field. But we needed more. We needed a preventive system that penetrated into every corner of the company, ensuring that the job was done right every time.

Our past business success proved to us that what we had been doing was correct. But it was costing us a fortune to be that good. What we didn't have was a quality system that penetrated every part of the business, or the understanding that the requirements set the minimum acceptable performance level. Today, conformance to requirements is not enough.

Part Two
IBM's Tactics for Excellence Improvement

Let me set the stage as IBM started the 1980s. If quality is defined as conformance to customer requirements, we were producing quality products by everyone's standards. Our products set the standard for the rest of the world. This was accomplished by the comprehensive use of the following:

- Design review.
- Design evaluation.
- In-process inspection.
- Failure analysis.
- Process qualification.
- Supplier surveys.
- Field and in-process reporting.
- Customer surveys.
- Operator training.
- Acceptable quality levels based on customer requirements.
- Basic belief by management that excellence is a way of life.
- Standard quality practices.

I like to think of our tactics for excellence improvement as a building that all of the corporate employees would like to live in. It is made up of five basic elements:

1. *The foundation is management action.* A house built on sand will wash away with the first major storm, but one built on bedrock will withstand the onslaught of time, erosion, and storm. It is imperative to build a firm foundation first, and to make sure that it is firmly anchored to bedrock. This means that the excellence process must be made part of the management system. It must become part of every management decision and become second nature to each manager. To many people, this will require a major change in business personality. Superficial support will give superficial results that soon will be washed away, like a sand castle built too close to the water's edge. John Akers, when discussing IBM's improvement process, said, "It must begin at the top and it will not happen any other way."

2. *The framework is process controls.* Process controls are the framework that supports the house and gives it strength. Bringing the process under control before starting the high production stage is the key to a preventive system of management.

3. *The siding is systems controls.* Systems controls are the siding that keeps the wind of unexpected problems and the heat of firefighting out of the home making it a comfortable place to live. Excellence (error-free performance) can be obtained only by focusing the improvement process on the system. Employees work *in* the system; management must work *on* the system. Only limited, short range improvement can be made by working with the employ-

7

ees. What management must do is focus on the business systems that limit the commitment to excellence employees want to give. Excellence can be accomplished only by paying meticulous attention to every small detail in the many business processes we use everyday.

4. *The roof is business associate (supplier) relations.* Effective business associate relations and involvement serve as the roof that protects the home from the rains of defective parts and the snows of missing supplier ship dates.

5. *The paint and the trim are total participation.* Until it is plastered, painted, and the molding put up, one cannot take pride in the home, nor is it ready for occupation. These seemingly small, final touches make all the difference. The same is true in the improvement process, for people *do* make a difference. Each manager and each employee must be an active participant in the excellence process and understand their roles in the corporation's search for error-free performance. They cannot just be aware of the process; they cannot just be involved in the process; they must be active participants in the process and be totally dedicated to making it a success. It's like having ham and eggs for breakfast. The farmer was responsible for the process that put the ham and eggs on the table. The chicken was involved in the process. But the pig was totally dedicated to it.

This book gives an overview of the tactics used by IBM to improve quality and productivity so that the goals for the 1980s could be reached. Many improvement tools were investigated, and the most effective ones were put into a toolbox that was made available to the total corporation. No effort was expended to impose the total list of tactics on every location, because we wanted each location to have ownership of its improvement process. Each group, division, and location selected the tools that best fit its needs, thus allowing each process to take on its own specific personality. This proved to be the most effective way of implementing the improvement process.

Tactic 1: Management Action

Quality improvement results from management action. The success of the improvement process is directly related to management's demonstrated commitment to it. This commitment requires both an internal and external change for many corporate leaders. Top management must become an active participant in the improvement process by communicating the importance of the process and, even more important, by actions that set the example for the entire corporation. In truth, the corporation becomes a mirror image of its president. At IBM the following management actions were taken.

Understanding

First we had to obtain a good understanding of where we stood, as related to our competition and our customers' expectations. We also felt it was imperative to have a good understanding of how well we were providing high-quality service between departments and to individuals inside the corporation. To define our internal weaknesses, corporate-wide opinion surveys were conducted, and a number of private personal interviews were conducted by personnel with all levels of management and employees who worked in all phases of the business.

To define how we were doing compared to our competition, we used data published by such groups as Reliability Plus and by our competition to project how rapidly they were changing, and how their product performance would improve in the future. To gain a better understanding of the latest quality improvement methods and techniques we sent managers to classes on quality improvement led by Philip B. Crosby, W. Edwards Deming, Kaoru Ishikawa, Joseph M. Juran, and other quality gurus around the world. In the early 1980s Crosby set up special classes restricted to IBM managers so that corporate problems could be discussed openly. Crosby's influence was prominent during the early implementation phase of the improvement process.

But as we got deeper into the improvement process, many consultants were invited to tour our plants and give lectures in the local areas, so that we could benefit from their varied experiences in implementing the improvement process. In some cases, consultants like Armand V. Feigenbaum were assigned to specific locations so that their approach to quality improvement could be evaluated firsthand for further application within IBM. This research provided management with sufficient information so that an improvement process uniquely tailored to the IBM environment could be planned and implemented. The process was a blend of all the leading consultants' concepts plus some unique IBM methods.

Direction

As quality improvement needs became understood, it was obvious that upper management needed to set the new direction for the corporation. To accomplish this the following steps were taken.

1. Management openly committed itself to the improvement philosophy by frequently publishing statements about the importance of quality, the error-free performance standard, and what was required to meet the continuing challenge for improvement. For example, Frank Cary, when he was chief operating officer of IBM, wrote, "Our reputation for quality is only as good as our last machine or our last customer call. As IBMers, none of us can be satisfied with a quality rating of 95 percent, 99 percent, or anything less than 100 percent. We should expect *all* our products to be defect-free."

2. A formal quality policy was released. It stated, "We will deliver defect-free, competitive products and services to our customers on time." This covers the big Q of quality — quality of cost, quality of schedule, and the conventional quality of output. In a comprehensive quality system, all three must exist. The product must be priced so that the customer can afford to buy it. It must be available to the customer when he or she needs to use it, and its performance must meet his or her expectations. Usually when a schedule is missed, it is caused by vendor or process problems. When cost targets are missed, it usually relates to a poor process design or a poor engineering design. In a complete quality system all three — quality, cost, and schedule — are brought under control.

The following are the definitions used by IBM to further define its corporate policy:

We — The corporation as a whole, and each employee as an individual.

Competitive — Provide the customer with more value for his or her investment than the competition can.

Customer — The next person that receives our output.

3. Two new corporate instructions were issued that had a major impact on the quality personality of the corporation. The first instruction simply told management that each new product must perform better than the product it will replace (current product) and the competition's product. As logical as this instruction sounds, implementation was difficult. Frequently, the new product would have twice the capabilities that the current unit had, and the current unit was operating much better than the specifications required (Figure 2). In the past, the specification requirements probably would have been the same as they were for the current unit, allowing the new unit to go through the normal early program improvement cycle. However, with the new requirement, the specification must be equal to or better than the actual performance of the current unit, which resulted in many designs being returned to the drawing board.

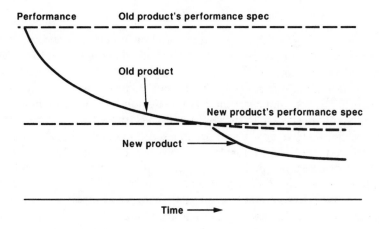

Figure 2. Product Performance versus Time

The second new corporate instruction was directed at establishing process control over critical business processes. The instruction required that each business process be assigned an owner, be certified, and be classified in one of five categories. A level 5 classification is a process that has not had the business process controls applied to it. A level 1 classification is a process where all ingredients of process control are in place and are working effectively to anticipate the customer's expectations. The process owner is responsible for organizing process improvement teams that will bring each process up to the first level. Typical activities classified as business processes are cost estimating, physical inventory management, supplier relations, engineering change implementation, product release, planning and scheduling, accounts receivable, inventory control, payroll, financial planning, fixed assets, and appropriations controls. (See Appendix I for a more detailed list.) This approach to continuous improvement will be discussed further in Tactic 3: Systems Controls, page 31.

4. We used the weekly and monthly newsletters to keep the importance of quality and the improvement gains visible to management and employees. This constant reminder of the importance of the improvement process is a valuable tool to gain and keep everyone's participation in the process.

5. Quality improvement plans became part of the annual strategic operational plan generated by each function. These plans included the resources required to implement the plan and the estimated savings that would be realized. Each area is held accountable for implementing these plans and meeting the committed savings. Many of the function managers developed five-year improvement plans to give continuity to the annual plan. Once each year, meetings are held off-site where individual function managers review their plans with the other function managers to ensure that they mesh with and meet the needs of the other functions they are servicing.

Reorganization

Management. To give quality the same level of importance and prestige as cost and schedule, it must be represented by individuals who are at equivalent levels and stature on the corporate organization chart as the people they measure (manufacturing, engineering, and finance). It was necessary to restructure our top management team to demonstrate IBM's commitment to quality and excellence. In the past, quality assurance had been represented at the corporate level by the vice president of manufacturing. As a result of the increased emphasis on quality, a corporate vice president of quality and divisional directors of quality were appointed. It is important to point out that they were named vice president or director of quality, not quality assurance. We selected quality because their role was much larger than the historical role of quality assurance. Their job relates to the total quality aspects of every unit within the corporation, not just the organizations that provide products and services for the external consumer.

Improvement Coordinators. To ensure that quality improvement was implemented effectively throughout the organization, high-level, respected managers were appointed as quality improvement coordinators (frequently called "quality czars") at the group, division, laboratory, and plant levels. Quality councils were formed at division and site levels, chaired by the division president or the highest level manager at each location, to provide the quality improvement coordinator with maximum support and active participation.

Business Controls Department. A business controls department was formed to focus on the process and systems that controlled and directed the business. It is responsible for helping management develop business systems that meet the changing needs of our rapidly growing business. This department conducts audits to determine if the controls in place meet today's requirements and if the requirements are being followed. Business controls' primary focus is in the white-collar areas (sales, accounting, personnel, industrial engineering, etc.), leaving the checks and balances for the products to the other quality assurance departments.

IBM Quality Institutes. After sampling the training programs available around the world, IBM decided that it would have to develop its own educational program. The program is designed to meet IBM's special needs, which are greatly impacted by the quality of the white-collar support groups (development engineering, product engineering, sales, accounting, finance, field services, quality assurance, etc.). As a result three IBM quality institutes were formed: one in Southbury, Connecticut, one in Belgium, and one in Indonesia. Their mission was to develop advanced quality training that uniquely met IBM's requirements. To accomplish this, professionals from all parts of IBM were assigned to a committee that prepared an outline of the training requirements and developed the curriculum.

Education

Executive Quality Education. As we entered the 1980s, we decided that it was time to undertake a massive project to provide quality-related education to the 400,000 IBM employees. In 1981 the IBM Quality Institute was formed to spearhead this project. Its first priority was to establish a quality awareness and quality tool seminar for the top 1,800 executives. A two-day program was prepared, and within an 18-month period, all of the top executives attended this course at one of the two quality institutes in the United States or Europe. See Table 1 for a course outline.

1. Quality Awareness
   - Fundamentals of quality
   - Worldwide quality challenge
   - Quality's impact on productivity
   - White-collar and blue-collar quality

2. Assessment of IBM's Quality Status
   - IBM's goals for the 1980s
   - Worldwide computer performance
   - Corporate quality direction
   - Corporate quality staff strategy

3. Life Cycle Quality Costs

4. IBM Managers' Role in Quality
   - Behavior change
   - Requirements setting
   - Measure of quality performance
   - Setting and improving quality goals

5. Overview of Quality Programs Across IBM
   - Software
   - Manufacturing
   - Marketing
   - Service
   - Administration

6. Quality Tools
   - Statistics
   - Managing breakthrough
   - Problem analysis methods
   - Special tools

Table 1. Typical Top Management Course Outline

Train the Trainers. The second priority was to establish a program to train middle managers at IBM locations around the world. To implement this expanded effort the quality institute developed a "train the trainer" program. The purpose of this program was to provide a minimum of one person from each IBM location with enough knowledge about quality methods, techniques, and tools, and how to apply them in both the blue- and white-collar areas so that he or she could return to the location to develop and implement the quality training program for middle and first-line managers. The quality institutes also developed and provided high-quality training aids that were used by the trainees, thereby greatly reducing the cost to prepare the education training package.

Middle Management Education. Middle management classes were conducted around the world. A typical program would be a 20-hour off-site meeting. At these meetings managers were introduced to problem-solving methods, process controls, statistical applications, and measurement methods. See Table 2 for a typical course outline. The classes were purposely mixed with students from multiple functions, because it was important that middle managers have a thorough understanding of the activities and plans being considered in their interfacing areas.

First-Line Management Education. Each function was then responsible for developing its own training program for line managers. The location trainers provided assistance, but the function's second- and third-level management team developed the content of the classes, the improvement examples, and the problem lists based on the interests and activities in which they were involved. This plan had two advantages: the function had ownership of the class, and each class was customized to meet the needs of its students.

Employee Quality Education. Once the management team had been trained, classes were made available to the employees, focusing on problem solving, simple process controls, and data presentation.

1. Introduction
   Agenda
   Preassignment questions
2. Quality Basics and Company Overview
3. Improvement Process Overview
4. Management Commitment
5. Quality Improvement Teams
6. Education and Training
7. Poor-Quality Costs
8. Measurements
9. Statistical Quality Control
10. Corrective Action
11. Participative Management
12. Excellence Planning
13. Goal Setting
14. Problem Solving
15. The Opportunity Cycle
16. The Business Process
17. Stopping Bureaucracy
18. Recognition
19. Team Quality Planning
    Case study

Table 2. Typical Course Outline for Middle Management Training

Focus Quality Education. The early quality education activities provided the basic information needed by managers and employees to start the improvement process, but it did not provide the depth of understanding needed in some functions. Therefore, the quality institute and the location quality coordinators developed additional quality classes directed at the specific needs of individual functions. Classes were provided on subjects such as advanced statistical analysis, experimental design, participative management, supplier interface, process control applications applied to the white-collar area, quality of design, etc. See Table 3 for typical classes.

Supplier Education. IBM's product quality is highly dependent on the quality of components and materials supplied to us by many suppliers throughout the world. As mentioned previously, we are so dependent on our suppliers that we frequently refer to them as business associates. For this reason, we focused our attention on involving our business associates in the quality improvement education process. Key individuals from most of our current business associates were invited to visit IBM plants where their parts were being used, so that they could see firsthand how important their products were to IBM. They also attended a one-day training class that explained the importance IBM was placing on quality and how this increased emphasis would affect them. To get other companies started on the improvement process, IBM sponsored improvement seminars in key cities around the world such as Singapore, Kuala Lumpur, Malaysia, and Bangkok, for their suppliers, potential suppliers, customers, and government officials. Table 4 shows the agenda for a typical seminar conducted by this author. In some cases, business associates attended longer training sessions at the quality institute to help them implement their own improvement process. In other cases, IBM training teams visited the supplier's facilities to provide quality improvement educational opportunities to large groups of their employees.

1. Business Process Management: Quality Focus on the Business Process	2 days
2. Teach the Teacher	8 days
3. Advanced Facilitator Training	4 days
4. Process Analysis Technique	2 days
5. Quality Professional Training for Managers	5 days
6. Quality Professional Training I	3 weeks
7. Quality Professional Training II	2 weeks
8. Software Development Quality	5 days
9. Information Systems Development Quality	5 days
10. Logic Design Quality	3 weeks
11. Mechanical Design Quality	4 weeks
12. Product Design Quality	2 weeks
13. System Design Quality	5 days
14. Statistical Process Control	5 days
15. Reliability	5 days
16. Personal Computer for Quality Control	4 days
17. Data Handling Statistics	3 days

Table 3. Typical Classes Presented at the IBM Quality Institute

Seminar on Quality — The Pursuit of Excellence

8:30-9:00 am	Registration
9:00-9:10 am	Film: *Pursuit of Excellence*
9:10-9:35 am	Opening Speech K. B. Low, General Manager, IBM Malaysia Speech YBM Tengku Razaleigh Hamzah, Minister of Trade and Industry
9:35-10:00 am	Coffee Break
10:00-10:15 am	Film: *Casablanca Nights*
10:15 am-12:10 pm	Presentation: The Improvement Process Jim Harrington
12:10-1:10 pm	Lunch
1:10-2:15 pm	Presentation: Excellence — The IBM Way Jim Harrington Film: *A Focus On Quality*
2:15-2:55 pm	Presentation: Cost of Poor Quality Jim Harrington
2:55-3:15 pm	Presentation: What is a Good Quality System? Jim Harrington
3:15-3:45 pm	Discussion
3:45-4:00 pm	Coffee Break
4:00-4:45 pm	Presentation: Interfacing with the IPO E. Pettine, Manager, International Procurement Office, IBM Singapore, Pte. Ltd. Film: *Commitment to Quality*
4:45-5:00 pm	Closing Discussion

Table 4. Typical Outline of Supplier/Customer Seminar on Quality

Share Meetings. Share meetings were scheduled every three months to keep the improvement process moving and to provide a continuous flow of new and proven ideas. The location quality czars were invited to attend these meetings and share their experiences related to the improvement process. This allowed the successes and failures to be aired openly so that the total corporation could benefit from each location's experiences. Frequently, outside consultants and representatives from other companies involved in the improvement process were invited to make presentations. As new outside training classes and tapes became available, a representative group from the share team would attend the class or review the tape, then report back to the total team on its content, interest, and applicability. The share meeting was purposely moved from location to location, thereby providing the team with living examples of the way different locations were using the improvement process. Often department improvement teams from the host location would present their activities. This not only provided the share members with good examples of the improvement process, but also gave recognition to the department improvement team members. These meetings proved to be very important to the total process, and many locations and/or business units that were half-hearted at first about the improvement process became believers based on the enthusiasm, examples, ideas, and data presented.

Measurement

Measurement is at the heart of any improvement process. If something cannot be measured, it cannot be improved. We found many IBM employees who believed that the quality of their output could not be measured. Managers in personnel, engineering, and finance, to mention a few, argued that their department's output was not measurable from a quality standpoint. Our answer to such arguments was, "If you can't measure the quality of output of your employees, how do you know who to fire and who to promote?" All output is measurable. One need only consider who will be using the output (the customer) and what the customer expects to receive (output specifications). Appendix II lists typical output measurements.

Mean-Time-to-Error. Today, mean-time-to-error is becoming more and more popular as a measurement term. This approach applies to every area of the corporation. It also simplifies the data system because only the number of errors needs to be recorded, ignoring the total number of tasks performed. Another way of using the same approach is to measure errors per time period. The evolution to error-free performance goes something like this:

1. Start by measuring errors per day. When the errors decrease to one per day, start to measure errors per week.
2. When the errors per week drop to one, start to measure errors per month.
3. When the errors per month decrease to one, start to measure errors per year.

Everyone is an error-free performer. The difference is in the duration between errors. We all need to strive to increase the interval between errors. It is a goal we can accept as a common goal.

Target Tightening. For years, we had managed by setting targets and expecting all targets to be met. Activities that met the target were good; activities that did not were in trouble, and the situation had to be corrected. Targets were used to focus management attention on problems. Areas and products that were meeting their targets were forgotten, and the effort was redirected to other areas. With the new emphasis on continuous improvement, management's thinking had to change. It was no longer good enough just to meet targets unless the target was zero errors. As a result we started to use two types of targets (Figure 3). One was a quality-cost target that indicates the level set by customer requirements at the beginning of the project. Cost estimates and product profits are based on that level, and all improvements over that point provide additional levels of excellence and/or cost reduction to the customer as well as increased earnings to the company and the employees. All managers are still expected to meet all quality-cost targets. The second type of target is called a challenge target. Challenge targets represent an improved performance level set by the department in its quest for continuing improvement. Some locations have made it a rule to reset all quality targets as soon as the target has been met for three consecutive measurement periods. It is very different to have an executive ask why a department has not reset its targets, rather than why it has not met some of its targets.

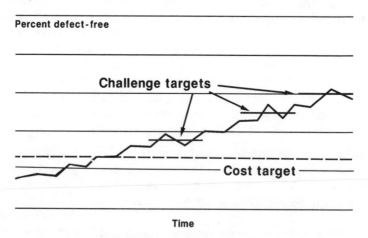

Figure 3. Challenge Targets

The important thing is not whether a challenge target is being met; the important thing is the trend that the measurement line is taking. It should always be improving; it should never be flat or going in a negative direction. If this occurs, then management should be concerned and start asking questions.

Management Focus on Measurement. Top-level corporate management conducts a detailed quality-improvement review, at least once a quarter, of the activities for all locations to ensure that there is a constant focus on the quality-improvement process. The division president and staff have a short, weekly review of the problem areas and a comprehensive quality review of all activities once a month. At the plant level, the plant manager has a comprehensive review of the major quality indicators once each week.

The Measurement Cycle. As you start the improvement process, you should measure Quality Activities. These include the number of people trained, the number of department improvement teams that are active, the number of control charts used in nonmanufacturing areas, etc. During the next stage, you should start to measure Quality Results, including the customer satisfaction level, poor-quality costs, the percent of suppliers at 100 percent conformance, etc. During the last phase, you should measure Business Results, including market share, return on investment, etc. All three measurements are required to sustain the proper focus on continuous improvement.

Poor-Quality Cost. IBM first started its poor-quality cost (PQC) system in 1963, working closely with Stanford University. The process operated effectively for about eight years and then was dropped. As we started the quality improvement process, we believed that PQC would be the best measure of improvement, so a minimum standard PQC system was implemented by the financial group at all locations. On a quarterly basis the vice president of quality prepares a corporate PQC overview report that is presented to the president of IBM. It was surprising how this old idea came back into favor with upper management (most of them were not involved with it in the 1960s), providing managers with a new reason to correct their quality problems.

Recognition

It was necessary to restructure the award system to show management's commitment to quality improvement and to be sure that the desired performance was rewarded. To accomplish this, a cash award structure was developed for individuals and groups who made major contributions to the improvement process, or did an outstanding job of preventing errors from occurring. For lesser contributions, trophies and plaques were given. The weekly and monthly location newsletter also provided recognition of improvement activities. Frequently, quality improvement teams were invited to the plant manager's staff meeting to present the results of a successful improvement activity. This had five advantages:

1. It provided recognition to members of the group.
2. It provided the employees with direct contact and exposure to top management.
3. It sharpened the group's presentation skills.
4. It reinforced management commitment and interest in the improvement process.
5. It provided a means by which good ideas could be shared.

Tactic 2: Process Controls

Process controls provide the siding that keeps the wind and cold out of your corporation. The best way to ensure that you are going to provide good output to your customers is to gain a thorough understanding of your process before you start to ship your output to your customers. Once you understand your process capabilities, strengths, and weaknesses, a knowledgeable control panel can be implemented to minimize appraisal and failure cost. This is accomplished in two ways: process qualification and statistical process controls.

Process Management

Process Qualification. Process qualification provides the corporation with a stable, understood process that meets customer expectations. This is the starting point for your continuous improvement process. It is the stake in the ground that sets your minimum standards for the future. This is the environment that makes maximum use of process control techniques. Why is process control management such a strong tool to assist in the continuing improvement portion of the product cycle? Process control management has the following five advantages:

1. It identifies trends before errors occur.
2. It identifies favorable process shifts so that management can evaluate their causes and take steps to make them a permanent part of the business system.
3. It allows undesirable output to be removed at the earliest point in the process.
4. It provides a means to ensure overall process utilization.
5. It provides a means that identifies when and where suboptimization occurs.

As John R. Opel said when he was president of IBM, "We have to do things right the first time. Each stage in the process must be defect-free output."

Process Ownership. In most companies the manufacturing process is extremely complex and has a number of support areas all driven by their own and often conflicting objectives. Purchasing is trying to buy parts at the lowest price. Production control is trying to keep inventories down. Quality assurance is perceived as not caring about cost so long as the end item is the very best. Manufacturing managers are worried about effective use of their employees. Manufacturing engineering tries to improve first-time yields. The subassembly areas want to meet ship schedules to the final assembly and test areas at the very lowest cost. Product engineering is working on the next product and doesn't have time to help out on the old products. With all these conflicting objectives it is easy to see why suboptimization often occurs and how time and effort can be lost, debating which objective will prevail. To offset this conflict someone should be assigned the responsibility for the effectiveness and performance of the total process. As a result we have assigned program managers who have total accountability for a product type in the United States and overseas. For example, we manufacture the same products in California, Germany, and Japan, and one program manager has the total responsibility for the products at all three locations. The manager's team expends a great deal of effort to ensure that all three locations are shipping products with equivalent quality and performance and that the best methods and ideas are implemented at all locations. There is one best way, and all locations should be using it. We call this process commonality, and it works to make the output of all locations better.

Process Control Engineering. Management and key employees must be trained in process control engineering methods to implement an effective process control system. This training should include the following:

- How to define inputs.
- How to document work activity and determine logical work flow.
- How to define measurable customer expectations.
- How to establish measurement and feedback systems.
- How to qualify processes.
- How to use statistical process control methods.
- How to collect and evaluate data.

Process Qualification

At IBM, a great deal of emphasis has been placed on process qualification before we ship products to our customers. Errors corrected before first customer ship incur only a small fraction of the cost that would have been incurred later on, and they don't affect the customer. To ensure that the process reaches a stable state, the processes evolve through the following four levels of qualification which provide management with a high degree of confidence that the output will meet customer expectations.

Level I   — Qualification of the Development Design
Level II — Qualification of the Model Build Process
Level III — Qualification of the Initial Manufacturing Process
Level IV — Qualification of the Final Automated Manufacturing Process

A typical component manufacturing cycle will evolve through the following stages (Figure 4):

- A theory or concept is proven by development laboratories that produce crude models using complex laboratory equipment.
- A model pilot line is prepared to manufacture larger quantities of product for internal evaluation.
- The production line is established to produce finished product for the customer.

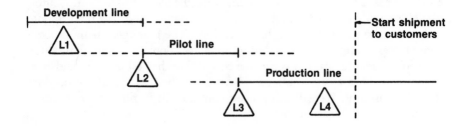

Figure 4. Component Process Development Cycle

Levels I and II qualifications evaluate the manufacturability of the proposed engineering design. Levels III and IV qualifications focus on equipment certification, process capability studies, process windows, operator training, first-time yield, throughput yield, and end-product performance under stress conditions.

Process qualification is the heart of a company's prevention activities. Eighty percent of all engineering support activities (both quality and manufacturing engineering) should be directed at the early-entry phase of the product cycle. If this phase is done correctly, the rest of the product cycle will run smoothly.

Certification applies to a single operation or piece of equipment. When an acceptable level of confidence has been developed that proves the operation and/or equipment is producing products to specification when the documentation is followed, that item is then certified. Qualification is granted to a complete process consisting of many connected operations that have already been individually certified and whose combined output is able to produce high-quality products consistently. For a process to be qualified, each of the operations and all the equipment used in the process must be certified. In addition, the process must have demonstrated that it can repeatedly produce high-quality products or services that meet customer expectations.

In the manufacturing process areas, normally quality assurance is responsible for process qualification. In the nonmanufacturing areas, process qualification is the responsibility of the process improvement team and probably will not go through all the stages indicated in this example. In the illustrated example there are four separate qualification levels. Level I qualification evaluates the acceptability of the development process. It takes place in the development laboratory during the early phases of the program. In its evaluation it is important to establish some basic controls, collect pertinent data, and study manufacturability without interfering with the creative nature of the work environment.

Level II qualification is directed at the pilot process used to produce products for internal evaluation and specification preparation. It is important to have a good understanding of the pilot process before these critical internal tests are run and specifications prepared. The purposes of Level II qualification are to:

- Characterize hardware and processes used to submit products to engineering evaluations.
- Provide a controlled environment for assessment of process and performance parameters.
- Establish a data base for the manufacturing process.
- Ensure that the process is ready to be transferred from development engineering to manufacturing.
- Provide management with a risk assessment of manufacturability and schedule integrity.

After establishing a controlled development pilot line, the next step in the process is to design and build a manufacturing production facility.

Experimental lots must give way to mass production quantities, and hours of processing time must be reduced to a minimum to meet cost targets. The manual development process must give way to automation. The equipment must undergo major changes to allow manufacturing operators to replace highly skilled development technicians. These activities present a major set of challenges to manufacturing engineering that result in drastic differences in the facilities. Manufacturability, as well as the ability to meet committed yields, must be demonstrated for the first time. Level III qualification evaluates this new process to ensure that it meets both customer and company expectations.

In most cases Level III qualification applies to a single-stream product line that has limited ability to produce customer-shippable products. Once this line is established the process continues to expand, adding equipment to the manufacturing facilities. Hard tooling replaces soft tooling and automation moves into the manufacturing process. During this growth period, sufficient quantities of products are made to allow extremes of specified test equipment settings and vendor material variations to be evaluated through process capability studies. This is the environment where we find ourselves as we start to perform Level IV qualification.

Up to this point the process qualification activities have been directed at bringing the process under control before we start shipping products to our customers. When Level IV is granted, the floodgates are opened so every control must be in place before this point in the program is reached. The purpose of Level IV is to characterize the process to ensure that it is under control and has the ability to produce output that consistently meets customer expectations.

Qualification Activities. To get a better understanding of process qualification at each of the four levels, let's take a look at the three major activities that go on during a typical qualification study:

- Certifying each operation in the process.
- Processing qualification lots.
- Conducting comprehensive program audits.

Certification. Certification activity looks at four facets of each operation: documentation, test and process equipment, operational requirements, and output acceptability.

Documentation. Documentation is an important part of any process as it allows the experience and knowledge of previous activities to be transmitted to the individual currently performing the job. An activity left to chance always has the potential for producing errors. Frequently companies build up a bureaucracy around paperwork; the intent is good, but the implementation is poor. Good documentation is short, to the point, and reflects optimum methods of performing the job. Most important of all, it must be easy to understand. During the certification evaluation of documentation, the following questions are asked:

- Are all the documents necessary to perform the job available to the person who is performing it?
- Are the documents released and properly controlled through the correct sign-off loops?
- Are the data collection systems in place and working?
- Is the software controlled and well documented? In today's companies, engineering change level control over software is important.
- Are training programs adequate and documented?
- Are process equipment, reliability, availability, and serviceability requirements documented?
- Is proper traceability in place?
- Are proper security plans in place?
- Is there any chance that the operators may misinterpret the documentation?

Test and process equipment. Test equipment and process equipment can have a major impact on the quality and productivity of the area. Certification of the equipment determines if it is capable of doing the assigned job and being maintained properly. Some of the activities appropriate to this facet of certification are as follows:

- Accuracy and repeatability assessments, including operator verification, must be conducted.
- Calibration and preventive maintenance procedures and intervals must be verified as correct.
- The equipment must be evaluated from a safety standpoint.
- Adequate controls must be established for standards.

- Correlation studies between similar pieces of equipment and other locations should be conducted.
- Long-term drift studies need to be performed.
- Extremes of settings must be evaluated and taken into consideration when the process window is established.

Operational requirements. At this point in the certification activities, we are evaluating the system, not the employees. It is the system that causes errors and places individuals in a position where they cannot fulfill their God-given right of error-free performance. Some of the things evaluated here are the following:

- Are the employees trained and do they believe the training was adequate?
- Are the proper controls in place on the environment that surrounds the activity (e.g., temperature, humidity, noise level, contamination, lighting).
- Are shelf-life controls in place?
- Have in-process and shipping containers been evaluated?
- Are the employees getting feedback on their performance?
- Has statistical process control been applied to the operation if it is appropriate?
- Are specifications for the job clear, or can they be left to interpretation?
- Have in-process and customer-failure analysis systems been established?
- Are controls in place so experiments do not confuse the normal product activities?
- Have process capability studies been completed on all major operations?
- Is the process statistically under control at all key operations?

Output acceptability. The first three facets of the certification activity have all been building up to the point where we can ensure output acceptability on a continuous basis. The following activities and requirements are typical for evaluating output acceptability for certification:

- Projected yields for this point in the program must be met for three consecutive months.
- In-process quality levels must be met for eight consecutive weeks.
- Daily going-rate projections must be met and must be following the learning curve.
- Returned part procedures must be in place to ensure that parts come back for failure analysis from higher level assembly.
- Stress tests must be in place to evaluate product reliability at the lowest possible level.
- Sources for all critical parts and materials must be established and approved.
- Component traceability must be established.
- Parts and components can only be purchased from approved suppliers.
- Supplier parts controls must be in place and followed.
- Correlation studies between suppliers and receiving inspection must be completed.

- Suppliers should be able to produce products that meet all the engineering specifications.
- Only high-volume suppliers can be used. Suppliers qualified at Level III must have sufficient process capabilities to meet volume demand at peak first-year production.

Qualification Lots. Once each activity in the process has been certified, qualification lots are processed to measure the effectiveness of the process design. The purpose of a qualification lot is to evaluate the continuity of the total process, to measure process yields, and to identify process volume limitations under controlled conditions. A typical Level III process qualification design experiment would be conducted over a five-week period; one of the five weeks would be used to measure equipment capacity and throughput volumes. A minimum of five separate lots would be processed through the manufacturing operations.

Independent Process Audit. The next step in the qualification process is a detailed process audit. The audit team is headed by the process improvement team chairman and consists of an independent group not assigned to the process. For example, you might have representatives from product engineering, development engineering, manufacturing engineering, product assurance, and sales. This audit team would evaluate the process status in the following areas:

- Have product and process manufacturability been proven?
- Is the complete process documented and understood?
- Has the design considered and corrected problems existing in similar products?
- Does the performance specification represent an improvement in reliability and quality performance compared to products that it will replace?
- Is the program schedule reasonable and does it have committed headcount and equipment funding from all supporting areas?
- Are there any major technical exposures to the program or the supporting technologies?
- Have the certification and qualification activities been implemented as required and have all major exposures been highlighted?
- Is the measurement and feedback system in place and working effectively?
- Does the end product meet customer expectations?

As soon as the audit team members have completed their assessment, they will meet with the process improvement team and report their findings. These findings will be documented subsequently in an audit team report. The process improvement team will generate corrective actions to solve each of the problems described in this report.

Measurements. The key to an effective process is to have good measurements that reflect customer expectation and/or have the ability to detect changes in the process before defective products are produced. Typical process control charts are:

- P charts — Percent defective.
- PM charts — Number of defectives.
- C charts — Number of defects in sample.
- U charts — Number of defects per unit.
- $\overline{X}$ and R charts — Average and range of variables data.

Histograms, percent defective, defects per unit, surveys, customer reports, and parts per million/parts per billion are typical measurements that are effective quality indicators when process control charts cannot be used.

Precontrols (Rainbow Control Charts). When small preproduction or production lots are involved, precontrol is the answer for IBM. This reduces the data necessary to establish the process control limits. Very simply, precontrol divides the specification tolerance in half and the operators are instructed to stay within the reduced specification, called the green operating range (Figure 5). When a measurement falls beyond the reduced limits into the yellow operating range, a second sample is taken. If it also falls outside the green operating range, the operator readjusts the equipment. When a measurement falls outside the specification limits into the red operating range, the operation is stopped.

Ground Rules for Using Precontrol Charts

1 Set up job. Evaluate five consecutive items.
  - If all five are in green range, setup is good.
  - If any are outside green range, redo setup and/or adjust.

2 At preset intervals, sample two consecutive items.
  - If first item is in green range, continue activity.
    (Do not evaluate second sample.)
  - If first item is in yellow range, evaluate second item.
    If second item is in green range, continue activity.
    If second item is also in yellow range, correct problem
    and redo step 1.
  - If first sample is in red range, correct problem
    and redo step 1.

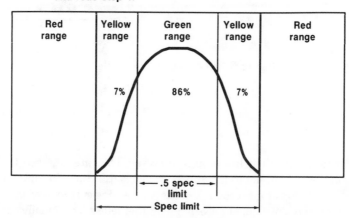

Figure 5. Precontrol Chart (Rainbow Control Chart)

Precontrol is based on the assumption that the normal distribution of a measured output can be kept within its specification limits by adjusting the controllable parts of the process. Precontrol also assumes that the process and engineering specifications are compatible. In this case, 86 percent of the output will be within the green operating range.

Process Control Cycle

The process control cycle starts with an unstable process that has many fluctuations (Figure 6). As the swings are understood and corrected, the process becomes more stable and eventually reaches a controlled state. Now you are ready to implement changes to the process that will lower the average defect level or narrow the population distribution. The product is not ready for manufacturing until the six sigma limits of the product output are well within the customer's expectations when the measurement error is taken into consideration.

Improvement Opportunities

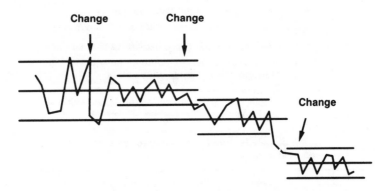

Figure 6. Process Control Cycle

Control charts have their limitations, however. They are designed to detect changes in the process, not to measure customer satisfaction. So control charts, although good, do not meet today's needs of satisfying customer expectations. What is needed is a new set of measurements that reflects continually changing customer expectations. These are the measurements that really count.

Tactic 3: Systems Controls (Business Process)

The key to improvement is focusing the improvement activity on the system, not on the employee. A system that is altered to prevent future problems from occurring is corrected forever. At best, a person can be instructed to do something right for as long as he or she performs the task; but without changing the training process, the next employee has a high probability of making the same error. The system, not the people, is the key to zero errors. Large corporations throughout the world have learned the value of detailed job descriptions, well-documented process flows, and extensive training instruction in the manufacturing areas to minimize the exposure to errors. This same concept applies equally well in the white-collar areas if you think of them as processes, for in reality that's what most white-collar activities are.

What is a Process? IBM defines a process as "a group of logically related tasks that when performed, utilize the resources of the business to produce definitive results." To put it more simply, a process can be defined as a series of activities that takes an input, adds value to it, and produces an output to a customer. It can apply to the activities of one person or a number of people, one department or a number of departments, or one function or a number of functions. Most activities are processes and should be controlled in much the same manner that the manufacturing process is controlled. It doesn't make any difference where the activities are conducted — in manufacturing, accounting, sales, engineering, personnel, anywhere. John Opel said, "Everyone in IBM has customers, either inside or outside the company, who use the output of his or her job. Only if each person strives for and achieves defect-free work, can we reach our objective of superior quality."

Any process that is repeatable can have process controls applied to it. Our experience indicates that over 75 percent of the white-collar activities are candidates for process control. We refer to our white-collar processes as business processes.

IBM released a corporate instruction that required the critical business processes throughout the corporation to have process control methods integrated into them. Appendix I lists typical activities that were classified as critical business processes.

Process Ownership. One of the first things we realized was that organizations were organized vertically, and that the business processes flowed horizontally. This caused overlaps and voids between organizations. When we went to the different organizations that were involved in a typical business process and asked who is responsible for the total process, we found out that no one felt responsible, or that many people felt they were responsible (and when you had more than one individual responsible, you really had no one responsible). So as in the manufacturing process, it was necessary to assign one individual the responsibility for each major business process at each location.

Process Classification. The process owner evaluated the present process and classified it into one of five categories. If a business process was classified as a five, that meant it did not have process controls applied to it. Business processes that were classified at the first level had total process controls, fully implemented and effectively operating (Table 5). The process owner was responsible for organizing process improvement teams and upgrading each process to the first level classification.

This is not as easy as it may look. None of our critical business processes are classified at the fifth level, but none of them have progressed to the point that we can classify them as level one processes.

Process Improvement Teams. To accomplish their goals, the process improvement teams:

- Define the limits of the process.
- Define the customer and his or her requirements.
- Perform an activity analysis at the task level.
- Flowchart the current process.
- Qualify the process.
- Collect current procedures and analyze their interactions.
- Develop a list of problems and prioritize them.
- Develop preventive action plans.
- Establish measurements, control systems, and feedback procedures.
- Develop input and output specifications.
- Use the quality ring for continuous improvement (Figure 15).

5	The process, as currently practiced, is ineffective; major exposures exist requiring expeditious, corrective actions. The basics of process quality management are not in place.
4	The process, as currently practiced, has some operational and/or control weaknesses that require corrective action, but the resulting exposures are containable, and the weaknesses can be corrected in the near future. The basics of process quality management are in place.
3	The process, as currently practiced, is effective and no significant operational effectiveness or control exposures exist. Substantial quality-improvement activity is in progress. Error-free criteria have been established.
2	In addition to category 3 requirements, major improvements (including simplification) have been made to the process with tangible and measurable results realized. Business direction is evaluated with resulting process changes anticipated and committed.
1	In addition to category 2 requirements, the outputs of the process are assessed by the owner and the auditor, from the customer's viewpoint, as being substantially error-free (i.e., to the level the process can reasonably deliver), and no significant operational inefficiencies are anticipated.

Table 5. Process-Assessment Criteria

The key to improved business processes is simplification. Unfortunately, the approach most frequently used is complication. Often you can save as much from simplifying the process as you can by computerizing it. You need to look at each activity and ask the following questions:

- Why is it being done?
- Does it have to be done?
- What will happen if it is not done?
- How long does it hold up the flow? How much time elapsed from the time the last operation was completed until the activity being studied is completed? Often it is simply a signature holding up the business process for three to five days.
- What is the total cost for the activity and how does it compare to the value added?
- What can be done to eliminate the need for the activity? If you are signing a document because you cannot trust another manager to make a good decision, what can be done to train or inform the first manager so that he or she can be trusted to make the right decision?

Bureaucracy is strangling American and world business! For example, we have a problem and put in additional checks to screen out the problem until it is corrected, then forget to take the screening activity out. Or you discover someone taking advantage of the system so you add additional approvals to keep the situation from ever repeating itself, without questioning the net savings to the corporation. Someone wants a report and to satisfy our customers, we modify our reporting system to have it on his or her desk every Monday at 8:00 am, even though it means a computer operator comes in at 5:00 am just to make this special run. Then two years later, and two managers later, the report is still arriving on the same desk at 8:00 am on Monday, even though it is not used. In fact, the manager who is receiving it thinks it is a standard report that goes out to many people and never cancels the report, even though it is not useful. Yes, to get the best results, simplify — then computerize.

IBM mounted a special attack directed at cutting through and stamping out the layers of bureaucracy that had become an integral part of the IBM system. Many locations assigned special groups to spearhead activities to reduce or stamp out the mire of bureaucracy that was slowing down IBM's movement. The "Stamp Out Bureaucracy" program implemented at IBM Brazil is a good example of what can be accomplished. Major savings and streamlining took place in almost every area of IBM Brazil's operations, greatly increasing productivity and job satisfaction, and reducing costs.

Tools for Business Process Improvement

There are a number of tools that will help the process improvement team bring the business process under control. Some of them are Performance Evaluation and Review Technique (PERT) charting, business systems planning (BSP), process analysis techniques (PAT), structured analysis/design (SA/SD), department activity analysis (DAA), and value control (VC).

These analytical tools are designed to use in the many different environments. Such methods as DAA are designed to be used at the department level. PERT and BSP are best utilized at high levels in complex systems that cross functions. Others such as PAT are effective at the department level and across department levels.

Performance Evaluation and Review Technique. PERT is a systematic way of analyzing a total process. Each task in the process defines its output, input, and timing. These data are then manipulated in a computer to define critical paths and develop interprocess dependencies.

Business Systems Planning. BSP was first developed to help the information systems (I/S) departments examine both short- and long-term requirements. It has since been expanded to encompass far more than just I/S requirements. Today BSP is best suited to the large process levels or to the many white-collar processes that are heavily dependent on I/S. Typical areas where BSP can be applied readily are purchase billings, real estate, construction, and publications. BSP requires a team of four to seven people led by an executive-level manager working full time for eight to 10 weeks. These people interview primarily executive-level managers to define the business requirements of the process being studied. They develop a model of the information flow through the process and scrutinize the related work elements. This analysis leads to improved data flow and restructured work elements. BSP has several advantages: the methodology is well evaluated, DP background is not required, software is available (ISNOD), consultants can help implement the program, and the process is completely documented and the methods well understood. Disadvantages to BSP include the need for interviewing skills and one-week training. Also it is a full-time assignment for eight to 10 weeks for four to seven people.

Process Analysis Technique. PAT is a tool designed to examine how work flows between areas. It helps the team to dissect, analyze, and reconstruct the process being studied, in an organized manner. This system requires two full-time people (a consultant and a lead coordinator) for four weeks. The process consists of the following 13 steps:

1. Trigger
   - Decision to do a PAT
   - Why the PAT is to be performed
2. Selection of process
   - Manageable for analysis
   - If more than one area will undergo analysis, start with the one that appears to have the most problems
3. Document of understanding
   - Details of analysis
   - Contract for analysis
4. Preliminary PAT
   - Meeting with involved management
   - Management describes process
   - Management names key people at task level
5. Schedule interviews
   - People at the task level
6. Information meeting
   - PAT explained to people at the task level
   - Five analysis questions presented
7. PAT interviews
   - People at the task level
   - Interview via five analysis questions
   - Current process documented in PAT flow
8. Analysis of business process
   - Scrubbing of tasks
   - Identification of tasks to be kept, eliminated, moved, or altered
9. Alternative solutions
   - Development of revised process
10. Identify dependencies
    - Intersecting processes
    - Sources of support
11. Test alternatives
    - Prototyping
    - Identification of additions, deletions, and corrections
12. Adjust alternatives
    - Final flow developed
13. Decision to implement
    - Decision
    - Method of implementation

This method relies heavily on the knowledge of the employees who are involved with the process to define the process and its problems, and to suggest improvements. The advantages of this method are that it requires only two full-time people for four weeks, no special background is required, the technology is generic, and it is a proven method.

Requirement of a consultant and the fact that only poor documentation is available today to support the method are the disadvantages.

Structured Analysis and Structured Design (SA/SD). The techniques of SA/SD are well documented in several books; one is *Structured Systems Analysis,* by Chris Gane and Trish Sarson (Prentice-Hall). Very simply, structured analysis provides a systematic way of communicating customer requirements from the customer to an analyst and from the analyst to the systems developer or designer. It helps define what is required of the system. To accomplish this it uses models as functional specifications, plus the techniques of mini-specifications, data dictionaries, data flow diagrams, data structure, and access diagrams.

Structured design is directed at how to organize a solution to satisfy the customer's requirements. Given a well-defined set of customer requirements, it will provide a solution that is flexible and maintainable. Using a logical model plus a set of objectives, it develops a system that fulfills the requirements. This is sometimes called top-down development. Typical tools used are data flow diagrams, data dictionaries, Nassi-Scheiderman diagrams, quality evaluation criteria, structure charts, and Jackson data structure diagrams.

Department Activity Analysis. DAA is discussed in **Tactic 5: Total Participation,** page 45, in the section on department improvement teams. As the name implies, it is most commonly used at the department level.

Value Control (VC). VC was developed by Armin R. Tietze, a senior engineer at IBM San Jose. It is a method of restructuring thinking patterns to help us see cost and quality dependencies. The procedure starts by defining perfection for any activity. Any expenditure over this ideal situation is then defined as imperfection. The level of imperfection is then reduced by either improving the existing system or changing the system to make it more effective, or both.

The data shown in Figures 7 and 8 were provided by Tietze. It shows that in a development department of eight people, only 1.85 people are defining the process and product which is the desired output from a development department. Support activities, such as coordinating, exploring, evaluating, and understanding occupy the other 6.15 people (Figure 7). Of the 1.85 people defining the process and product, 1.39 are doing administrative work such as documentation, communication, and similar "bureaucratic" activities. Less than 0.5 of the eight people's time is actually spent doing the main job (perfection).

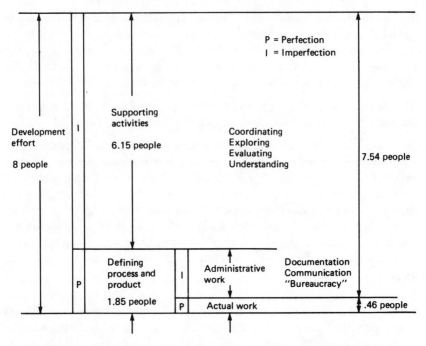

Figure 7. Process Development Department with Eight People

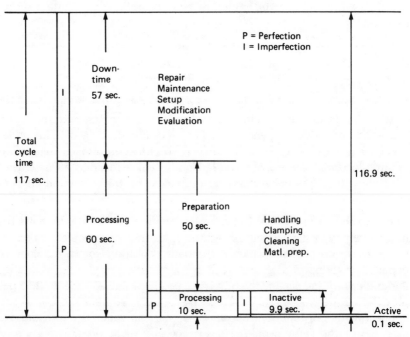

Figure 8. Process Cycle Time

Figure 8 analyzes a buffing operation on the disk manufacturing line. In this case dividing the number of good parts processed into the number of hours available gives a total cycle time of 117 seconds per disk. Of that time, the equipment is not being used for 57 seconds due to repair, maintenance, set-up, modification, or evaluation; thus level 1 perfection is 60 seconds per disk. Of that 60 seconds, 50 seconds are used in preparation activities such as disk handling, clamping, cleaning, and material preparation. The buffing tape is in contact with the disk for only 10 seconds (level 2 perfection) of the total 60 seconds of processing time. Of the 10 seconds that the buffing tape is in contact with the entire disk, it is only on each area of the disk for 0.1 second. Perfection time in this case is 0.1 seconds; imperfection time is 116.9 seconds.

Opportunity Cycle

Each problem encountered is really an opportunity for you or the team to contribute to the success of your company, not a bottleneck interfering with the process. Using the opportunity cycle (Figure 9), the process improvement team makes a list of problems keeping it from meeting customers' expectations and/or adding additional, unnecessary costs to the process.

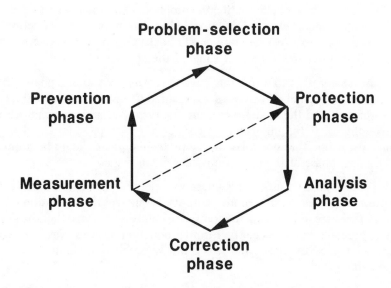

Figure 9. The Opportunity Cycle

1. *Problem-selection phase.* The problems are then prioritized, based on customer impact and financial returns. Two pareto analysis curves are needed, one based on customer impact, the other based on cost of the problem to you and your customer. These two curves are used to select priority problems.

2. *Protection phase.* Frequently it is necessary to implement a temporary corrective action to prevent the error from reaching your customer until the preventive action can take effect. This might take the form of increased inspections or talks with employees alerting them to the error and asking them to take extra care.

3. *Analysis phase.* Once the problems are prioritized and the customer is protected from the problem, we can start the analysis stage of the opportunity cycle. During this stage, the symptoms are analyzed to determine the true disease. Typical tools used to define the true disease are failure analysis, design experiments, and error duplication. When the true disease is known you can make the problem appear and disappear at will.

4. *Correction phase.* The next phase of the opportunity cycle is the correction phase. During this phase, the individuals who can correct the problems are identified and they accept responsibility for developing an action plan that will prevent the problem from ever recurring, or at a minimum, developing a plan that will reduce the frequency of occurrence to an acceptable level. All corrective action plans must include an estimate of how effective the corrective action will be at eliminating the problem. Once the plan is prepared, the corrective action plan will be modeled to determine if it is effective. If it is not, a new corrective action plan will be developed. If it is effective, we are ready to progress to the next stage of the opportunity cycle.

5. *Measurement phase.* During this stage the corrective action is phased into the process and actual effectiveness is measured to ensure that the estimated improvements and the model results are not degraded under normal working conditions. If the implemented corrective action provides the desired results, the action that was taken in the protection phase should be stopped and the last phase of the opportunity cycle can begin.

6. *Prevention phase.* During this phase we apply what we have learned in the previous phases to other similar outputs throughout the corporation. We also use the same information to modify the control systems so that the same type of problem will not occur on future products or services. Now the team goes back to the first phase and selects a new problem to eliminate. It is a never-ending cycle.

Systems Controls. The company's operations and their effectiveness are controlled by systems operating procedures that explain how the business activities of the company will be performed. If the systems are ignored, it is easy for these systems to become obsolete, cumbersome, or bureaucratic. As the process improvement teams do their work, they are responsible for updating the operating procedures to keep them current with present practices and goals. Remember that the job is not done until the paperwork is finished. The controlling paperwork must be changed to make the improvements a permanent part of the company's operations.

Typical System Improvement. The following is only one of the many examples of improvement that occurred because we focused on our business processes. The procedure used to process nonstandard customer requests for products (special bid process) was defined as one of our critical business processes because it accounted for a large amount of IBM's revenue. This process consists of a proposal generated by marketing, an engineering design, an estimate of the manufacturing costs, and finally, a firm proposal presented to the customer. The process was taking an average of 90 days. The sales force considered this to be its most important process, but considered the process to be unwieldy due to the many delays. The sales force believed that if the cycle time could be reduced, the possibility of making a sale would be greatly improved. Thirty days was believed to be a reasonable target.

The division vice president was designated as the process owner, and a process improvement team was formed to systematically remove anything that inhibited the free flow of the special bid process. The number of decision points was reduced. Greater authority was given to the sales team, and an automated data collection and approval system was installed. Within 24 months the average cycle time was reduced from 90 to 15 days. That is not the important part of the savings, however. Certainly the effort required to process a special bid was greatly reduced, but the big return came from increased business. Bid closure rates jumped from 20 percent to 65 percent, a major increase in revenue.

Tactic 4: Supplier Relations

No corporation or company is an island; all are dependent to some degree on other companies that provide them with material, energy, components, resources, etc. At IBM, our product quality is greatly dependent upon the quality of the goods and services provided by our suppliers. As mentioned earlier, our suppliers play such a critical role in our business strategy that frequently we refer to them as our business associates. Our employees are our most important asset, but our suppliers run a close second. To allow us to progress toward our error-free goal, we found it was necessary to develop a team relationship and establish common goals with these important business associates. We changed many of our old beliefs about supplier interface to accomplish this.

Let me set the stage. When we started this new team relationship with our suppliers, a typical location scoreboard would read:

Suppliers at 100 percent conformance for one year   78 percent
Average lots rejected per month   240
Percent of rejected lots returned to suppliers   53 percent

The manufacturing process was frequently stopped (at least once a week) to remove defective components. This caused high levels of rework, shortage of parts, and special processes to allow products to continuously flow around missing components. The process was continuously changing, not to improve the process, but because it was necessary to work around the many defective components that slipped through receiving inspection. It was a classic case study of how receiving inspection cannot be relied on to inspect quality into supplier parts. The key to obtaining high-quality components is working closely with the suppliers at their home locations, not trying to second guess them in receiving inspection.

Fewer Suppliers. For years we have required at least two suppliers for every part number, and more than two was preferable. This provided protection against strikes and catastrophic plant failures. A great deal of money was spent every year evaluating potential new suppliers to give us this additional protection. To keep the multiple suppliers active, purchase quantities were split between the many approved suppliers, with the larger quantity going to the supplier who had the lowest price. What this did was to fragment the orders, not allowing the suppliers to properly tool up their processes to get maximum productivity and quality, or obtain minimum cost. Today there is no requirement for dual sources. We rely on our sources acquired by our overseas facilities to help us out in case of emergency. This has led to a massive reduction in the number of suppliers used at IBM, allowing us to focus more resources to help improve the suppliers that are already producing high-quality output. It has also allowed us to reduce travel expense and manpower required to support suppliers.

Long-Term Contracts. For years IBM has been careful not to overcommit to suppliers. Most orders were placed on a month-to-month basis, and suppliers had little or no knowledge about how many follow-up orders they would receive. Again this had a negative effect on their tooling plans and many other financial considerations. Today we are using long-term contracts (maximum one year) for components and providing the suppliers with a no-commitment projection beyond the contract limits. This has allowed the suppliers to plan better, to improve their process, and to retain a higher-level work force.

Cost to Stock. Today, we are considering much more than purchase price when we make the decision to buy. The cost of a component is the cost to put it in stock, which includes receiving inspection, quality engineering support, and procurement manfacturing engineering support. The cost to stock is the cost factor that should be used in price considerations.

Design Reviews. There has been a continuous effort to involve suppliers early in the product cycle so that they can contribute to defining how their product is used and specified. This proved to be fruitful for everyone because no one knows the product better than the supplier who is manufacturing it. As a result supplier input is the key to ensuring the product's proper application and specification.

Supplier Process Control Management. We have helped many suppliers to implement process control management. A joint IBM/supplier team has flowcharted and documented their process. The team then works together to establish control points that will be used to collect process data. In some cases the suppliers send their process control data to IBM by computer before the parts are shipped. A few suppliers have sensors in their process that transmit the in-process data directly to IBM. IBM uses the flowcharts and documentation to audit the supplier's process to ensure that it has not changed.

Performance Feedback. Regular quality reports that measure their performance are sent to the suppliers. Each time a lot is rejected, the procurement quality engineer contacts the supplier to review the problem within four hours of the time the parts are rejected. This allows the supplier to take immediate action to minimize the cost impact and to correct the problems on the parts that are going through the line. Each time a lot is rejected, suppliers are required to document their preventive action and submit it to IBM. Although supplier feedback has, for the most part, been limited to performance in receiving inspection, efforts are under way to establish systems that will provide the suppliers with line and customer performance data.

Incentive and Penalty Programs. Poor quality costs IBM dollars. Good quality saves everyone dollars. It is only reasonable to expect suppliers to bear some of IBM's costs when they provide poor quality. The following are two typical examples of penalty programs that have been used by one IBM location with cable suppliers:

---

*This concept is not implemented in areas where it is prohibited by law.

43

- **Approach I**
  For defect levels of 0.0 to 0.2 percent, the supplier receives premium price; for defect levels of 0.21 to 0.3 percent, the supplier receives $2 less per cable; and for defect levels of 0.31 percent and above, the supplier receives $4 less per cable.
- **Approach II**
  In this case IBM was using source inspectors to accept parts at the supplier. When the source-inspection data shows defects are not being controlled by the supplier, the supplier reimburses IBM for the out-of-pocket costs for the source inspector's return trip to reevaluate rejected lots.

Is this concept effective? Just look at Figure 10. For years the average supplier defect rate had been running from 0.1 to 0.12 percent. Within two months of the time the penalty contracts were implemented, the defect level dropped to 0.04 percent and has remained close to that level since.

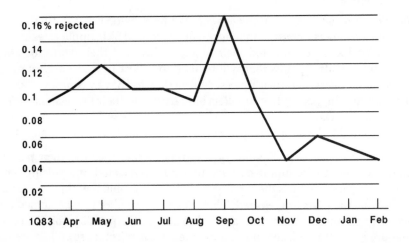

Figure 10. Supplier Defect Rate

Tactic 5: Total Participation

The final touch that makes the house a place you want to live in is the total participation of each member of the corporation in the improvement process. The company that limits the improvement process to the output delivered to the consumers outside of the company realizes only a small portion of the benefits of the improvement process. Often because the supporting organizations are not improving, the production employees become discouraged, believing they are being used as whipping boys while the rest of the company goes on providing the same old shabby output. As John Opel said when he was president of IBM, "Quality is not the exclusive province of engineering, manufacturing or, for that matter, services, marketing, or administration. Quality is truly everyone's job. Each function, each individual in IBM must assume the responsibility for a defect-free operation." This is the type of management direction that must come from the top and be repeated over and over again in many different ways so that managers and employees both know they are expected to become active participants in the improvement process. This is how you get that last 20 percent of employee-related errors out of the process — the 20 percent that sets your company apart from the run-of-the-mill.

Leadership Training

IBM has always sought out the employees to make them active partners in the business and allow them to make decisions that directly affect them. William Ouchi, in his book, *Theory Z* (Addison-Wesley Publishing Company, 1981), stated that IBM was more Japanese than Japan itself. But even with a history of consensus decision making and team building, we decided it was time to reexplore the tools used by effective leaders. (Notice I use the word *leaders*, not managers.)

Managers are appointed by management and with the stroke of the pen given accountability and responsibility to direct an activity, but management cannot *make* any individual a *leader*. Leadership is a learned skill that requires a great deal more than just technical confidence and/or job knowledge. All too often excellent employees are promoted to management because of superior job performance, but they have never learned to become good leaders. When this occurs, management has done the new manager, his or her employees, and the company an injustice. IBM so strongly believes in the need for managers to be leaders that they developed a dual ladder concept, allowing technical people who were not inclined to lead others to continue to develop and be promoted to pay levels equal to those of the management team.

It is imperative that management training be directed at leadership skills as well as business tools. To provide the required focus on leadership skills, managers were sent to a five-day class that highlighted techniques like participative management. At the same time, new management training programs were restructured to emphasize employee involvement in the decision-making process and the use of leadership skills.

Quality Improvement Teams

Following this training program, quality improvement teams were formed in each department. Every employee in the department was expected to participate in the department quality improvement team because each employee was expected to support the department improvement activity. The department manager served as the improvement team chairman. The improvement team started by preparing a department activity analysis.

Department Activity Analysis. To help each department understand its role in the quality movement, department improvement teams fill out department activity analysis forms (Figures 11, 12, 13, and 14). On the first form (Figure 11), the department improvement team describes the department's mission and lists its major activities. Each activity is then analyzed to determine who uses its output. The next form (Figure 12) defines who receives the department's output from each activity and how those customers measure quality. It is also used to record the quality level that the customer requires. Figure 13 shows the form used to analyze the value-added content the department contributes to the output from each activity. The department's effort is then divided into basic value-added work and poor-quality cost categories. Finally (Figure 14) each activity is analyzed to determine what inputs are required, where they come from, and what the department's quality requirements are for those inputs. Both the supplier and the customer sign off on the document, indicating their agreement. It is instructive and surprising when the customer is asked to sign off on the measurement of output quality. Frequently what the department thought was important is not what the customer considered important, but rather there is some other value that the customer considers most important. In some cases the output being supplied is not even needed and can be dropped. As a result of the department activity analysis, the department can now establish key quality indicators, called team improvement charts (TIC). Each department is encouraged to post a minimum of three TIC charts in a place where they are readily visible to the department members.

Department Activity Analysis

Function Name

Department Name | Department No.

Department Mission

List Major or all Activities/Tasks/Responsibilities of Department

Manager's/Preparer's Name | Date | Extension

Department Approval

_____   _____   _____
_____   _____   _____
_____   _____   _____
_____   _____   _____

Figure 11. Department Activity Analysis: Mission and Major Activities

Note: Use additional pages if more space is needed.

Activity	Department	Date	Prepared by

Output

What are the output requirements that you and your customer have agreed to?
-
-
-
-
-
-

What are the quality measurements that will show if your output meets requirements, and how will they be measured?
-
-
-
-
-

Customer Approval

_____   _____   _____
_____   _____   _____
_____   _____   _____
_____   _____   _____
_____   _____   _____

Figure 12. Department Activity Analysis: Customer Requirements

Note: Use additional pages if more space is needed.

Activity	Department	Date	Prepared by

Value Added — Work Accomplished in Dept.

Why do?

Value added

Impact if not done

How many hours/week are spent on this activity? _____ hrs/wk

These hours are either a cost of doing business, a poor-quality cost (PQC), or some combination of the two. How do you classify them?

       Business _____ hrs/wk
       PQC   _____ hrs/wk

Those hours that are PQC can be further classified into prevention, appraisal, and failure. What are they?

       Prevention _____ hrs/wk
       Appraisal   _____ hrs/wk
       Failure   _____ hrs/wk
       Total PQC _____ hrs/wk

Figure 13. Department Activity Analysis: Value-Added Analysis

Note: Use additional pages if more space is needed.

Activity	Department	Date	Prepared by

Input

What

From

What are the input requirements that you and your supplier have agreed to?
-
-
-
-
-

What quality measurements will show if supplier's output meets requirements?
-
-
-
-
-

Supplier Approval

_____ _____ _____
_____ _____ _____
_____ _____ _____
_____ _____ _____
_____ _____ _____

Figure 14. Department Activity Analysis: Supplier Analysis

Quality Improvement Team Education. The next phase of the department improvement team's activity is to attend a training program where they learned about data analysis, problem solving, sampling process control methods, and how to use the opportunity cycle. Typically a quality improvement team would receive training in most of the following areas:

- Brainstorming
- Nominal Group Technique
- Force Field Analysis
- Cause and Effect Diagrams
- Mind Maps
- Checksheets
- Graphs
- Histograms
- 80-20 Rule
- Pareto Diagrams
- Sampling
- Data Collection
- Data Arrangement
- pn Control Charts
- $\overline{X}$-R Control Charts
- Stratification
- Scatter Diagrams
- Presentation Skills
- Group Participation Skills

Quality Circles. On occasion, one of the items on the problem list has low priority for the department, but is a high-priority item for a subgroup of the department (two to four employees). In such a case, a quality circle is formed to attack the individual problem and report the results of its activities to the department improvement team. The quality circle is dissolved as soon as the individual problem is solved. In locations where quality improvement teams were not used, quality circles were used to meet individual needs.

The improvement team approach is favored over conventional quality circles because it requires the involvement of every employee. In addition, the improvement team approach provides everyone with a common vocabulary, problem-solving techniques, a better understanding of company and department goals, and a common understanding of who their customers are.

Continuous Improvement

In today's environment we can never stop improving. Too many people try to justify this week's pay based on last week's press clippings. The day you stop improving is the day you start to slip backward. The output that was excellent yesterday just meets the requirements today and probably will be inadequate tomorrow. We all need to strive for a goal that is just out of our reach. An objective of error-free performance fulfills this need. It is the pot of gold at the end of the rainbow. It is the Shangri-la just over the hill. If we try to jump over the moon, we will never make it, but we may set a new world record for high jumping, and that's not so bad. Thomas J. Watson, Sr., the first president of IBM, said, "It's better to aim at perfection and miss than to aim at imperfection and hit it."

In our quest for error-free performance we need to concentrate on continuous improvement. We use the quality ring (Figure 15) to help us continuously improve. We start by setting new targets, then develop a plan to meet the targets. We then implement the plan, adjusting it as necessary based on new data that become available. When we meet the target, we take time out to celebrate our accomplishments and reward ourselves, then we set new targets, develop new plans, implement the plans, reward ourselves, set new targets, develop new plans, implement the plans, and reward ourselves. It is a continuous cycle — around and around and around the quality ring, continuously improving, never satisfied with where we are, and always getting closer and closer to error-free performance.

Figure 15. The Quality Ring

Management by Objectives

This is an old technique that has proven to be an effective tool for improving an individual's productivity and quality of output. Each year the manager and the employee develop a performance plan (contract) for the coming year. Part of every individual's yearly performance plan is a measure of the quality of the individual's output. A typical manufacturing operator would have 40 percent of the performance based on the quality of his or her output, 40 percent on productivity, and 20 percent on how effectively he or she interfaces with the other department members and works as a team member. How well the individual performs to this plan determines how much money he or she will earn over the coming year.

Feedback

A key factor in keeping people actively participating in the improvement process is to provide them with data that measures how well they are performing. This feedback loop can be accomplished in many ways. Some of them are:

1. *Performance evaluations.* At least once every three months, every manager sits down with each employee to analyze how well the employee is performing, based on his or her performance plan. At least once a year a formal performance review is conducted and documented; it is used to set the employee's salary for the following year.

2. *Customer meetings.* There is no more effective way of providing feedback and understanding of your customer's expectations than to invite your internal customers to attend a department quality improvement team meeting. Visiting your customer's work area to discuss the problems they are having with your output and to point out areas where they feel you have improved is also effective.

3. *Team improvement charts.* TIC charts are posted in prominent places in the department and kept up-to-date. Three or four charts showing measurement of the department's output provide an effective way of keeping the employee's attention focused on the improvement process.

4. *Self-evaluations.* The department members know the real story. They know that things are getting better; they also are their own worst critics. Periodically, ask all the department members to evaluate the progress that the department is making by filling out confidential evaluation forms. Have one of the team members summarize the forms and report back to the group.

5. *Opinion surveys.* Every 18 months the total population of IBM participates in an opinion survey that is strictly confidential. Each employee evaluates and rates such things as:

- How good a job is your manager doing?
- How well are your skills being used?
- How would you rate IBM as a place to work?
- How much help is your manager giving you in developing your potential?
- How good is the communication upward?
- Does management put more emphasis on schedule than on quality?

The survey forms are turned in to personnel. A computer program is used to analyze the results and provide the management team with a statistical analysis. To ensure confidentiality, data are never broken down to units with fewer than eight employees.

Part Three
Quality Improvement Process Cycle

Some corporations are very successful in implementing improvement processes; others are dismal failures. Those that were unsuccessful had one thing in common: a short implementation cycle of one or two years. Those that were most successful all had long-term implementation cycles of three or more years so that the improvement process was absorbed into the management system, rather than being applied as a blanket that can be thrown off easily when things get hot. Our improvement process started in 1980 and we are still implementing our business-process concept. There is no use starting an improvement process if management is not willing to make a long-term commitment to it. The plan for implementation must also provide for thorough management training and observed management participation before employees can be asked to improve. After all, only management can correct 80 percent of the problems impacting the company's performance.

IBM has been a successful company by anyone's measurement standards. When we started the improvement process, surveys conducted by non-IBM companies verified that IBM products were good and getting better and that our quality was setting worldwide standards. We had a large backlog of orders that assured us continuous growth. Revenues and profits were at a record high and the *Fortune* magazine poll showed IBM as the most admired corporation. But that is only part of the story. We felt we were good, but it was costing us a fortune to be that good. Preventive costs were running between two and seven percent of revenue, appraisal costs between four and 10 percent, failure costs between nine and 23 percent. Putting it all together, poor-quality costs varied from 15 to 40 percent of revenue, depending on the complexity and maturity of the product being analyzed and the management system being used (Figure 16).

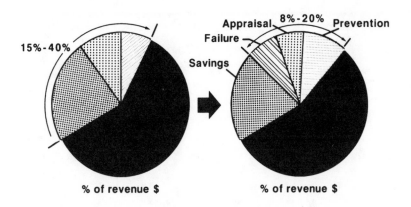

Figure 16. IBM's 1983 Poor-Quality Costs

We realized that by reducing poor-quality costs, we could substantially reduce manufacturing costs, increase customer satisfaction, decrease internal frustration, and decrease product costs while increasing profits. As John Akers said, "We are responding to problems as they surface instead of doing everything possible to prevent them from occurring in the first place. More important than cost, some of these problems were hurting our hard-earned reputation." Based on the success of the first three years of our improvement process, Mr. Akers reported in a meeting with the security analysts in Boca Raton, Florida, "We believe quality improvement will reduce quality cost by 50 percent over the coming years. For us that translates into billions of dollars of additional profit potential and quality leadership in our industry."

Taking 1983 as an example, IBM's earnings before taxes were nearly 10 billion dollars:

1983 earnings before taxes	$9,940 million
Five percent quality improvement savings	$1,048 million
Increased earnings	10.5 percent

Sales were more than $23 billion. If we reduced poor-quality cost by a mere five percent of sales, we would realize a savings of $1.2 billion. Assuming we spent $116 million on the improvement process during the year, we would still receive a net savings of over $1 billion. That represents a 10 percent increase in earnings, without considering the additional products made available to our customers as a result of not wasting our resources by redoing work that has already been completed. When this is taken into consideration, we would realize almost a 20 percent increase in revenue from reducing poor-quality costs by five percent. All this without adding one new person, or any floor space or increase in equipment.

Part Four
Results

The results we have seen to date have far exceeded our expectations. Total costs have been reduced significantly. Profits have increased. Customer satisfaction has reached an all-time high. Employees are more satisfied with the work they are doing. We are seeing a much higher return on investments. When discussing IBM's quality improvement process in 1985, John Akers stated, "IBM's revenue has grown 53 percent over the last three years. . .while the number of employees has grown only nine percent. During the same period, our profit growth has exceeded our revenue growth at an increasing rate — and we know our quality efforts have had something to do with these trends."

The ultimate measure of success is how well we are tracking to the four goals we set for ourselves in 1979. The following goals and data cover the period of 1980 to 1983:

- *To grow with the industry.* Revenue growth rate is up at a compound rate of 15.3 percent and revenue dollars up by 53 percent (Figure 17).

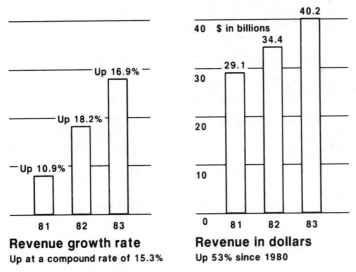

To Grow with the Industry

Revenue growth rate
Up at a compound rate of 15.3%

Revenue in dollars
Up 53% since 1980

Figure 17. Revenue Growth

- *To exhibit product leadership across our entire product line,* to excel in technology, value, and quality. One of the best measurements we have of this goal is improved price performance for our customer. Over the initial period of our improvement process, computing cost was down on all product lines, led by the high end of our products where a 33 percent improvement was recorded. Typical of these advances was the direct access storage devices that almost doubled their density from seven million to 12 million bits per square inch (Figure 18).

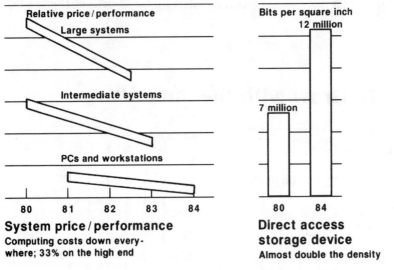

Figure 18. Price per Performance

- *To be the most efficient in everything we do,* to be the low-cost producer, the low-cost seller, the low-cost servicer, the low-cost administrator. Between 1980 and 1983 return on assets increased by 20 percent. Revenue per employee was $109,400 in 1983, up 42 percent over 1980; by 1984 it had increased to $120,000 (Figure 19).

To Be the Most Efficient in Everything We Do —
to Be the Low-Cost Producer, the Low-Cost Seller,
the Low-Cost Servicer, the Low-Cost Administrator

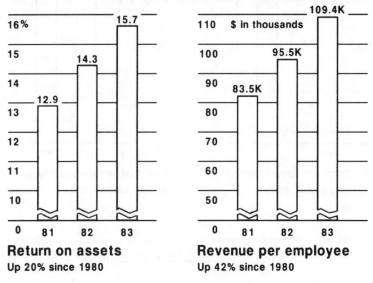

Return on assets
Up 20% since 1980

Revenue per employee
Up 42% since 1980

Figure 19. Return on Investment

- *To sustain our profitability, which funds our growth.* Return on investment is up 19 percent and net earnings, before taxes, have increased 70 percent in the first three years of our improvement activities (Figure 20).

To Sustain our Profitability, which Funds our Growth

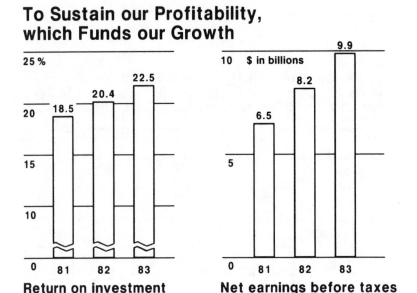

Return on investment
Up 19% since 1980

Net earnings before taxes
Up 70% since 1980

Figure 20. Sustained Profitability

Certainly these data prove that we are meeting and exceeding the goals that we set for ourselves. Improvements of this magnitude do not result from individual single actions. They are an accumulation of many small improvements that add up to increasing the effectiveness of every business process.

Summary

The improvement process at IBM has reduced costs, increased productivity, increased profits, improved customer relations, aided our employees in taking more pride in their work, and helped IBM remain the number one data processing company in the world.

We have made progress, but we have only started down the long road to a preventive management system and our goal of error-free performance. As John Akers stated, "Impressive as all these improvements are, we really believe our greatest gains are ahead of us. As the quality movement matures, quality improvement is becoming an integral part of the way we do business. The interesting thing is that it feeds on itself. . . it is becoming business as usual."

Our solutions may not fit your situation but by sharing our experiences, I hope I will help stimulate your thinking. And maybe, just maybe, some of the things we have found to be successful can be applied to your company.

There is an old World War I song that says, "It's a long way to Tipperary." Well, it's an even longer way to error-free performance. All employees at IBM have picked up their improvement tools and are marching down that long road, stamping out errors as they go along. Why don't you join us in our march against waste and help your company, your nation, and yourself to a better way of life? Let's all accept a personal goal of error-free performance. Let's do the job right *every* time.

Part Five
Examples of Improvement

The improvement process is made up of a myriad of small changes and improvements. It is not like an electric light that you turn on and instantaneously the room is flooded with illumination. It is more like the dawn, where first a few rays of light dare to stretch across the black expanse of sky, then slowly the night is transformed into the glorious day as trillions of individual rays of light cover every corner of our existence, lighting our way and warming our land. The following are some of the rays of light that helped us see and understand the value of continuous improvement.

New Product Quality Requirements. One of the most effective drivers of our quality movement was a very simple corporate instruction that stated, "Before we announce a new product, our product must be better in terms of quality and availability than the product it replaces, or our competition's product." A typical new IBM design would provide our customers with improved performance at the same or reduced cost. Frequently a new machine was able to replace two old ones. Using this as a basis, we reasoned that the mean time between failures for the new machine should be targeted for the same level as the original specification for the old machine, providing the customers with a 100 percent improvement in reliability. The fallacy in this reasoning is simple. When the old machine was replaced, it was performing much better than its specifications, and that is the performance data customers are using to compare their replacement products.

Figure 21 shows the repair action for one of our data processors. Curve A is the historical curve for the current product. You will note that at general availability time (GA), the product was meeting engineering requirements and customer expectations, but over the next 30 months there was a major improvement in the product's performance. Its new data processor, curve B, based on the new corporate instructions, was required to perform better than unit A is performing when the first new unit is delivered to a regular customer — GA. In the past, the new product, which was much faster and had many more capabilities, would have had a GA performance target close to that of the current unit at GA. But as you can see by the graph, the current product went through a learning curve which has driven the quality level to a point that has never been achieved before. Our new products have to be much better than they were in the past to comply with our new corporate specifications which require that "before we announce a new product, our product must be better in terms of quality and availability than the product it replaces, or our competition's product." This specification requires that the new product's start point is at the bottom of the current product's learning curve for both performance and quality. You can see how much better product B was at GA than the current product A. The design did have a major impact on accomplishing this feat, but the design itself does not account for the drastic change. These results were achieved through the use of the many ingredients of the improvement process and by management paying meticulous attention to each and every detail.

Data Processor

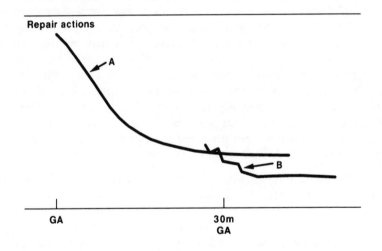

Figure 21. Action for Data Processor

The New Design Cycle. Figure 22 is the curve for one of our new midsize data processors. These four curves represent four different design proposals that were prepared over a two-and-a-half-year period. In curve 1, the proposed design had a better projected performance than the current unit had at general availability, but its initial performance was not equal to the current product's present performance. That was not good enough, for our new corporate instruction required it to be better at its initial ship point than the current unit was performing when the new product was first shipped to a customer. As a result it took us four iterations, over a two-and-a-half-year period, to meet the quality objectives required by the corporate instruction. Then, and only then, was it announced.

Data Processor B

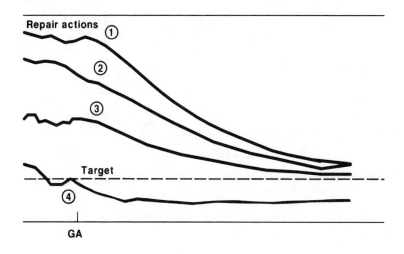

Figure 22. Data Processor B

Large Data Processors (308X Family): Installation and Early Life Error-Free Performance. One of the best ways to measure manufacturing quality is to measure error-free performance during installation and warranty periods. Figure 23 tracks the improvement in error-free performance of our largest processor, the 308X series product. For years, we had believed that a defect-free installation of our large computers was impossible because of their complexity, buildup of reliability figures, storage-life failure rates, and shipping problems. But once we focused on error-free installation, the process immediately began to improve. Today the 308X series is seven times better than its predecessor product in terms of defect-free installation. As a direct impact of this improved quality, installation time has been cut by a factor of three. Performance after installation has also taken on totally new characteristics. The present 308X has more than twice the mean time to failure as its predecessor product.

308X Processor
Defect-free at installation

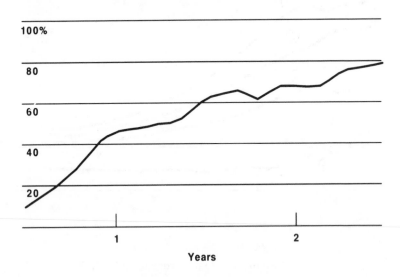

Figure 23. 308X Processor: Defect-Free at Installation

Selectric Typewriter Warranty Performance. We had a tendency to forget our old products. If our customers were happy with them and they were meeting our engineering specifications, we set them aside and went to work on something new. But the impact the improvement process can have on old products is very significant. The Selectric typewriter, using 25-year-old technology, is an excellent example of what can be accomplished (Figure 24). When we started the improvement process, we looked at the Selectric typewriter and realized it was performing well, but not well enough. We thought it could be better. We set targets for ourselves that many people felt were unrealistic, but in just a two-year period, there was a fourfold improvement in performance at installation, and a 40 percent improvement in customer performance was realized in just an 18-month period. After four years there was an eightfold improvement in repair actions.

Selectric Typewriter

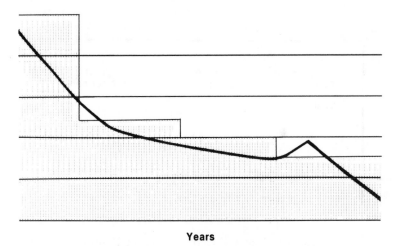

Years

40% improved customer performance in 18 months

Figure 24. Selectric Typewriter Performance

Flat Cable Improvement. For over 20 years IBM has been connecting their electronic circuits together, using flat ribbon cables. Present usage rate is about two million per year. This is a well-known technology that was moved out to vendors years ago. IBM supplied the bulk cables and the connectors, while vendors did the assembly. For years price, schedule, and quality requirements had been met, but met at a tremendous cost. Massive amounts of rework had been built into the product estimate and treated as normal expected yield. A young industrial engineer from IBM Raleigh, North Carolina, mustered up enough interest from management so that a team was assigned to attack the high defect rates present in the plant and at the suppliers. Methodically the team conducted a detailed analysis of each type of defect. When they were confident that they truly understood what was causing the problem, they took steps to eliminate it, emphasizing changing the process to improve first-time yields. The results were very effective, as Table 6 indicates. As the result of this one effort alone, our annual savings were $5 million.

	Before	After
Scrap	.94¢ per cable	.28¢ per cable
Quality		
Rework	25%	4%
Final test	12%	1.2%

Table 6. Flat Cable Improvement Results

Mother-Board Improvement. The mother-board is made up of a group of connectors and circuitry that connects and supports the logic cards. Again this was an old technology that had been around for over 20 years. Everyone believed that it had already been improved as much as possible. But in reality when we looked closely at the process, it was a typical example of the philosophy that says if you need more product, push more into the beginning of the process. As this old product was transferred once again to a new manufacturing site, the management team questioned why it was necessary to put up with low yields throughout the process. As a result the defect rate dropped from 10 to 2.1 percent in less than three years (Figure 25). In fact, the product reached such a high-quality level that our Poughkeepsie, New York plant, which installs the mother-board in our mainframes, was able to stop receiving inspection of the part for the first time, saving $1 million a year. The reduced defect rate in the process itself resulted in savings of $1.5 million. A total of $2.5 million a year — a fantastic return on investment!

Board Pinning Process

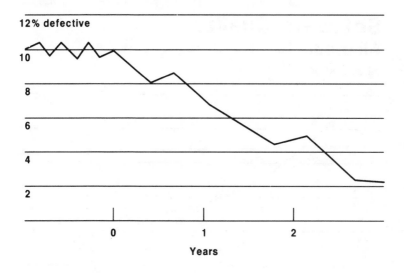

Figure 25. Board Pinning Process

Process Improvement Applied to Software. To measure the quality of software, we monitor defects per 1,000 lines of code (KLOC). Our thrust in the 1970s was not to reduce the created level of defect, but to focus on additional appraisal activities to reduce the number of defects shipped to the customer. These efforts were effective (Figure 26). For the period from 1976 to 1982 there was a threefold improvement, primarily as a result of more effective appraisal systems. However, the slope of the line was becoming flat and it was difficult and expensive to expand our appraisal system to capture the small defect levels that were "escaping" to our customers. As a result, we took a different approach and emphasized prevention rather than appraisal. We increased our focus on education, tool development, and process management. We set up our own institute on this science discipline. This new approach was first evident in 1983 when we announced three major software products whose error rate varied between two and three times lower than the average of the software products shipped during 1982, setting new world-class standards of excellence for software. Since 1983 we have continued to expand our emphasis on prevention. This has resulted in some major programs becoming error free, and was a major reason we won the NASA Excellence Award in 1987.

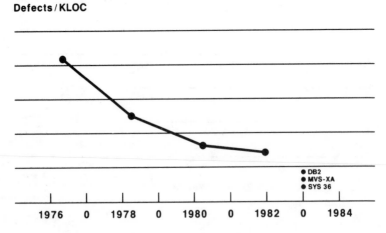

Figure 26. Software Quality on all Products

Process Improvement Applied to Accounting. When we began to apply the improvement process to the accounting area, things began to change. To get the process going, we decided that miscodes entered into the computer would be the measure of excellence. We were pleased to find that the data entry system was 98 percent accurate, but when we looked at the corporate accounting office, which was processing over a million codings per day, that two percent error rate translated into 20 to 30 thousand miscodes per day (Figure 27). Managers in the accounting area took responsibility for the entire process and things began to improve. They focused their attention and the skills of their employees on the entire process, setting up feedback systems to help improve the quality of the input data as well as increasing the accuracy of their activities. As a result the error rate dropped from two percent to 0.4 percent in a two-year period.

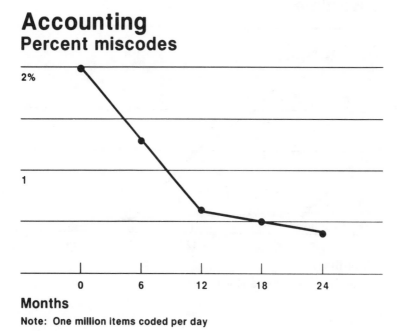

Figure 27. Accounting Miscodes

Process Improvement's Effect on Overtime in Accounting. Two percent mis-coded items took a disproportional part of the department's time because they were difficult to detect and correct. As a result, the function was working about 45 percent overtime (Figure 28). Within 24 months from the time the improvement process was implemented, overtime dropped almost to zero — and, in fact, in many months it was zero. Another side effect of implementing the improvement process was a significant improvement in employee morale in the accounting function.

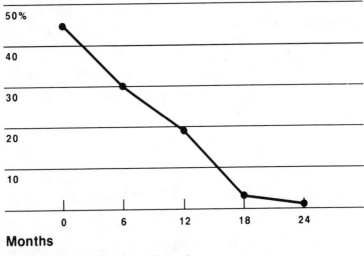

Accounting
Overtime

Months

Note: **One million items coded per day**

Figure 28. Accounting Overtime

Back-Order Improvements. The plant department that supplies field engineering with replacement parts decided to measure its quality based on the number of orders that were not filled in three days (back orders). They accepted total responsibility for the entire process and, as a result, soon found out how to reduce the number of back-ordered parts drastically. Within six months the number of back orders was cut 60 percent (Figure 29). The significance of this chart is that once the improvement gains were made, they were permanently implemented in the process, causing it to stabilize at that point and not degrade.

Parts for Field Engineering Division

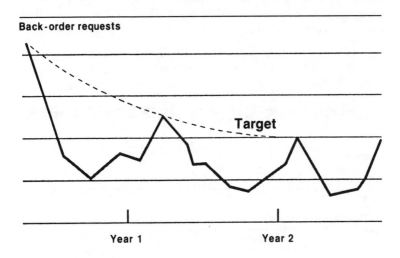

Figure 29. Parts for Field Engineering Division

Instrument Calibration Delinquency. Paperwork at the San Jose, California plant indicated that on an average 20 percent of the manufacturing equipment was out of calibration (Figure 30). The calibration department accepted responsibility for the entire process. Before that the calibration laboratory had sent out recall notices and left it up to the managers to return the instruments. Quality inspectors ran around the manufacturing floors checking calibration due dates and applying "Do Not Use" stickers whenever a delinquent piece of equipment was found, but this system was not effective. The calibration department's investigation uncovered two main problems: many of the overdue pieces of equipment were located in engineering laboratories and in many cases had not been used for years and, much of the delinquent equipment could not be found in the manufacturing department that received the delinquency notice. In most cases the equipment had been transferred to another department, but the calibration records had not been updated. To correct these problems unused equipment was turned in to calibration, reducing the need to buy new equipment. Also, the equipment inventory data base was linked to the calibration data base, eliminating the need for managers to notify two areas when they transferred equipment. As a result of our understanding the true disease, within six months instrument calibration delinquency dropped to zero percent.

Instrument Calibration Delinquency

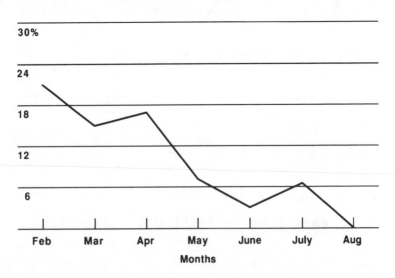

Figure 30. Instrument Calibration Delinquency

74

Rejected Lots Returned to Suppliers. For years when defective parts were rejected in receiving inspection, we tried our best to "off-spec" or rework them so that we could still use them. We reasoned that many of the rejected parts would present no functional problem even though they didn't meet the engineering specification. We reasoned that if we rejected the parts the supplier would repair them, and those costs would be reflected in future orders. So we tried to use them if possible, thinking we were saving money and keeping the line moving. But we were defeating ourselves. What we were really doing was telegraphing to our supplier that the specifications were not important. To turn the situation around we decided not to off-spec parts unless there was a dire emergency or unless engineering had agreed to modify the engineering specifications in the near future. As a result the percentage of defective parts rejected and returned to the supplier increased from 53.8 percent to 92 percent in four years. Now the suppliers know that we really do expect parts to meet print requirements.

Rejects Returned to Suppliers

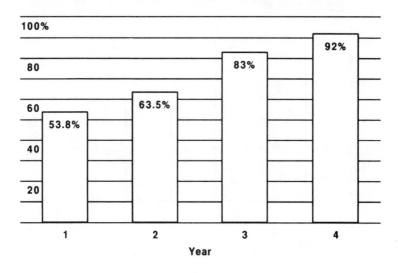

Figure 31. Rejects Returned to Suppliers

Receiving Inspection Rejections. As a result of a "no off-spec" policy, establishing close working relationships with fewer suppliers, and cleaning up our own act, receiving inspection rates plunged. As we began to work more closely with our suppliers, we soon realized that all the problems with supplier parts were not theirs, that IBM also had its share. Our instructions and specifications were not always clear, and at times suppliers received conflicting instructions from different sources within IBM. We decided to take the approach that if the problem was not obviously the suppliers, IBM would be responsible for it until there was enough data to fix the responsibility clearly. This, along with our other supplier improvement activities, caused the receiving inspection rejections at one of our plants to drop from an average of 232 to 64 lots per month over a four-year period, even though the quantity of parts sent to stock more than doubled during that same period (Figure 32).

Receiving Inspection Rejections

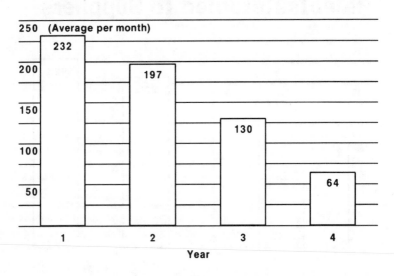

Figure 32. Receiving Inspection Rejections

Suppliers at 100 Percent Conformance. We believed that our supplier base was too large and decided to concentrate on a few of the very best suppliers to make them better, rather than trying to bring the whole supplier base up to an error-free performance level. As a result the number of active suppliers was greatly reduced, and the suppliers with a 100 percent lot-acceptance record for one full year jumped from 78.1 percent to 96.6 percent of the total supplier base in the San Jose process receiving inspection (Figure 33).

Suppliers at 100% Conformance

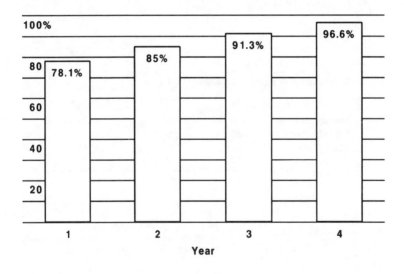

Figure 33. Supplier Conformance

Supplier Performance. In the 1970s we made the mistake of talking to our suppliers in terms of average quality levels (AQLs) when we should have been talking parts per million. If we had been talking parts per million in the 1970s, today we would be talking parts per billion defect rates. As we started our improvement process and increased our emphasis on suppliers providing defect-free parts, we were extremely surprised by the enthusiasm that we encountered in almost every situation. Suppliers were eager to work with us and readily accepted our help to improve their products. The results were extremely gratifying, as indicated by Table 7. In a three-year period, we saw improvements as high as 116 times. Integrated circuit (IC) linears typify the results that were achieved. In 1980 they were performing at 9,500 defects per million parts. In 1983 that was improved to 300, an improvement of 31.7 times. Our suppliers are still working with us to achieve defect-free products and today their products consistently surpass the results documented here. We are well on our way to a parts-per-billion defect rate supplier base.

Defect level in parts per million			
	1980	1983	Change
Transistors	2800	200	14x
Transformers	4200	100	42x
Capacitors	9300	80	116x
IC linears	9500	300	32x

Table 7. IBM Supplier Component Quality

Process Procurement Quality Assurance Excellence. Process procurement quality assurance is made up of receiving inspection and the procurement quality engineering group, who are responsible for all components and materials that enter the head/disk process at IBM San Jose. They established the number of times it was necessary to screen the manufacturing floor to remove bad supplier parts as their measure of excellence (Figure 34). In 1980 the manufacturing process was shut down seven times to remove bad supplier parts from the process. In 1983 the process was only interrupted once. The cumulative average for 1984, 1985, and 1986 was one process stoppage per 13 months. When you consider the thousands of lots that were processed, and the millions of decisions that were made by the entire quality assurance procurement group, that's close to error-free performance. How was it accomplished? Part of the credit has to go to the suppliers whose quality greatly improved during the period. Much of the credit also has to be given to the receiving inspection area, whose efficiency greatly improved as they implemented inspector certification by part number, improved workmanship standards, and identification and traceability of the actual parts handled.

Dept. 351 Head/Disk Procurement Quality
"The reject escapes"

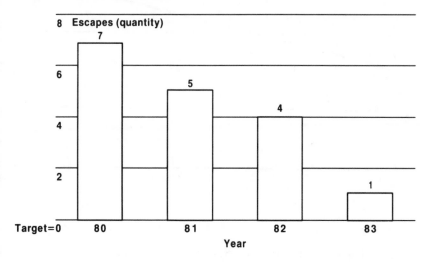

Figure 34. Department 351 Head/Disk Procurement Quality

Information Systems Performance. The data processing group decided to measure the excellence of its performance by the number of times reports were generated and delivered on schedule (Figure 35). In the past, input data, computer system downtime, and poor instructions from customers provided ample excuses not to have every report out on schedule. The department accepted the responsibility for the entire process, which included making sure the inputs were correct, instructions were thorough, and system problems were overcome. As a result the jobs delivered on schedule improved from 86 to 94 percent in a two-year period.

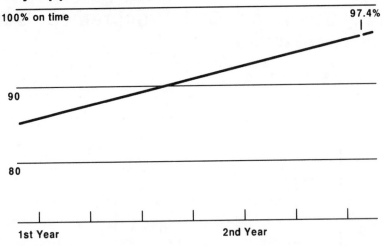

Information Systems Performance
Key applications

Figure 35. Information Systems Performance

Perpetual Inventory Improvement. Good data integrity is an essential part of any manufacturing process, but when the inventory records at our Raleigh, North Carolina plant were compared to the actual stock location, error rates as high as 30 percent were noted (Figure 36). Typical errors were wrong number of parts, wrong packaging, damaged packaging, and parts in wrong location. When the improvement process focused on resolving the differences between the inventory data and the actual stocked parts, the error rate improved to 1.7 percent — more than a 15-times improvement.

Perpetual Inventory
Error-rate improvement

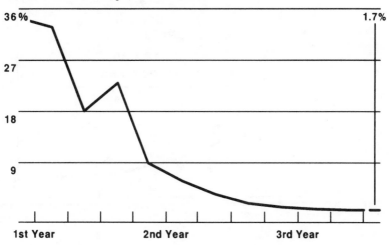

Figure 36. Perpetual Inventory

Late Payments. For many reasons not totally controllable by the financial department, late payments of billings ran about five percent, and we were processing over 18,000 bills per day (Figure 37). When we started treating the entire payment cycle as a process, within a two-year period the late payments dropped from five to 0.027 percent — a 200-times improvement.

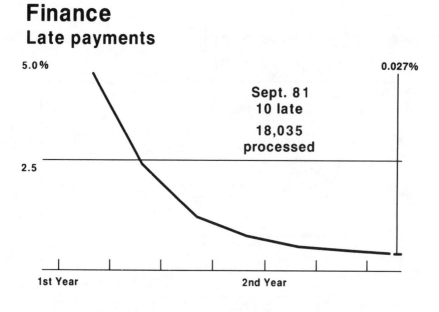

Figure 37. Financial Department Late Payments

Manufacturing Engineering Service (MES) Kits Error-Free Improvement. As feature changes or engineering changes are implemented at customer offices, MES kits are assembled to meet the individual configuration of the customer's equipment and its engineering change level. These kits consist of large quantities of parts, software, and special documentation. Prior to starting the improvement process one of our plants was averaging 91.7 percent error free. After the team focused its attention on the process, they were able to consistently ship kits that were better than 97 percent error free (Figure 38).

MES Kits
Defect-free improvement

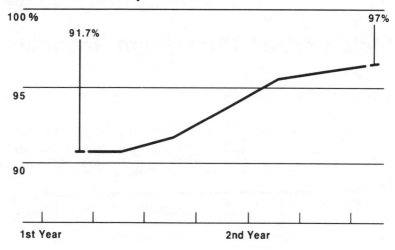

Figure 38. MES Kit Improvement

Field Service Cost/Point Reduction. One of the best ways to measure and improve field performance is by measuring service costs per dollar revenue (points). Figure 39 shows that IBM has been making excellent progress in reducing the cost of maintaining equipment in the field. In fact, over a six-year period there has been a 52 percent reduction in field service cost per point. But as the improvement process began to take shape, field service costs dropped even more rapidly, and the seven-year reduction had increased to 61 percent, well below the target we had set for ourselves. This trend has continued as shown by the following examples:

- The IBM personal computer had a fourfold warranty performance improvement in 16 months.
- The 3081 had a fivefold performance improvement in 24 months.
- Six major products from our plant in Havant, England, have only 18 percent of the problems they had in 1981.
- The 308X saw a sevenfold improvement in defect-free shipments and a twofold improvement in mean time between failures. Further, installation time is 33 percent of the previous system.

Field Service Cost / Point Reduction

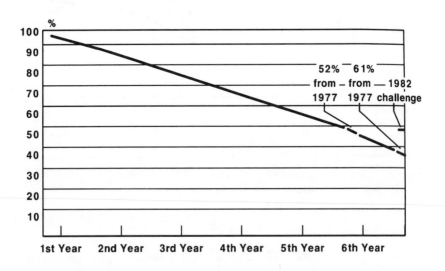

Figure 39. Field Service Cost/Point Reduction

Part Five

Individual Measurements. The methods covered in this book have worked for IBM around the world. For example, IBM Havant began to display graphs at the workstations showing how well the operators were performing. Figure 40 shows a summary of the assembly line for week 1 with each operator's initials clearly shown on the related bar graph. The second graph represents the same area just 15 weeks later. At that time 97 percent of the product was going through with no defects found. Not only did this technique reduce the number of defects but, even more important, it made the workers feel they owned the process and were responsible for it. With this feeling of ownership, peer pressure became an important factor in overall improvement. As important as peer pressure was, the feedback system provided the vehicle that allowed the improvement to occur.

Printer Final Inspection Results

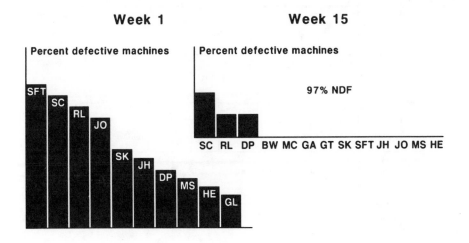

Figure 40. Printer Final Inspection Results

Improvement in Service Areas. Using IBM Havant as an example, Figure 41 shows how the number of manufacturing problems caused by the service area (parts supply, parts quality, DP services, and engineering support) decreased over a four-year period while the improvement process was being implemented.

Improvement in Service Areas

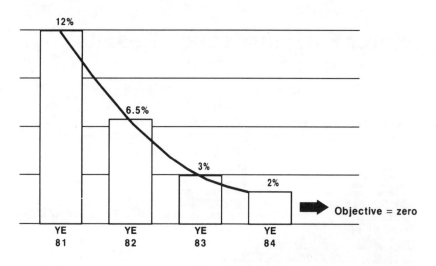

Figure 41. Improvement in Service Areas

Havant Product Field Performance. Figure 42 shows the improvements made in all products manufactured at IBM Havant in installation and first 90-day performance. This is the graph used by the plant to measure how successful they are at improving in the customer's environment. This graph is prominently displayed in the cafeteria, keeping the importance of customer performance readily in view of all employees. Many of the products produced at this plant have performed 100 percent defect free during installation and first 90 days for over a year. (Data on IBM Havant provided by P. K. Doran, Assistant Plant Manager, Havant Plant, IBM United Kingdom Ltd.)

Product Field Performance
Installation and first 90 days

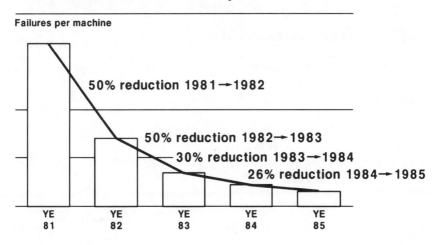

Figure 42. Product Field Performance

Product Comparisons. IBM manufactures the same products using compatible processes at two or three locations around the world. This allows friendly competition to develop between manufacturing sites, which results in improved performance for our customers. Figure 43 shows a comparison of the same intermediate processor's customer performance data, manufactured in Havant and Japan. Note that at the beginning of the 1980s, Havant's customer performance was not as good as Japan's, and Japan's product cut its repair action by about two-thirds in a four-year period. However, by the end of 1983, Havant was slightly better than Japan, and continued to improve into 1984. Now our Japanese plant personnel are coming to England to see what they are doing to be so good, and soon the lines will cross again. This shows that it is not the country's culture that makes the difference in quality; it is the control over the system and the management personalities that make the difference.

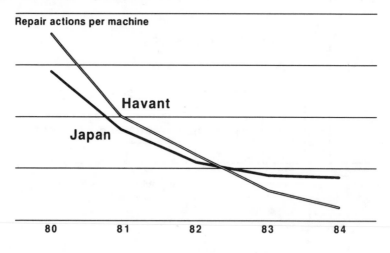

Product Comparisons
Intermediate processors

Figure 43. Product Comparisons

Participation. The emphasis on quality and training in problem solving had a major impact on our suggestion program. In a three-year period, the number of suggestions submitted increased by 48 percent, and the number of people participating in the suggestion program increased by 26 percent. The quality of the suggestions also improved, as the awards increased by 94 percent and the savings to IBM increased by 70 percent. We also conduct a program called *Cost-Effectiveness,* which allows an individual to claim savings for ideas that are directly related to his or her job and are not eligible for the suggestion program. As Figure 44 indicates, participation and total dollar savings also went up significantly in this program.

Participation

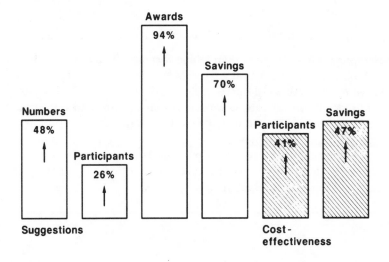

Figure 44. Participation

Morale Improvement. We have all heard that improved quality improves morale and that people want to do good work if you just give them a chance. IBM conducts an opinion survey of its entire 400,000 employees once every 18 months. This opinion survey is an anonymous evaluation of how the employees feel about their jobs, their manager, their progress, and their company. The survey is conducted by personnel and the raw data is fed directly into a computer for analysis. As a result there is absolutely no way the results of the survey can be traced back to an individual. Over the three-year period during the initial implementation of the improvement process, the following opinion survey gains were realized:

Job performance	22 percent
General satisfaction	12 percent
Skill utilization	14 percent
Personal development	13 percent
Communications — up	18 percent
Communications — down	16 percent
Job content	10 percent
Job involvement	6 percent
Morale index	9 percent

(Source: Corporate IBM Personnel Research Department)

The employees felt that their own job performance had improved by 22 percent, their general satisfaction with IBM had improved by 12 percent, their skills were being utilized 14 percent better, they were developing personally at a 13 percent greater rate, communications upward had improved by 18 percent, and communications downward by 16 percent. In every case there was a significant improvement in the morale questions in the opinion surveys; that is abnormally high based on past history. Allowing and helping our employees to do quality work and take pride in their output does improve morale and increase productivity.

Appendix I

Typical Business Processes Where Process Control Can Be Applied

Function	Process Name
Development	Records Management
	Acoustics Control Design
	Advanced Communication Development
	Cable Component Design
	Reliability Management
	Cost Target
	Design Test
	Design/Material Review
	Document Review
	High-Level Design Specification
	Industrial Design
	Interdivisional Liaison
	Logic Design and Verification
	Component Qualification
	Power System Design
	Product Management
	Product Publication
	Release
	System-Level Product Design
	System Reliability and Serviceability (RAS)
	System Requirements
	Tool Design
	User/System Interface Design
	Competitive Analysis
	Design Systems Support
	Engineering Operations
	Information Development
	Interconnect Planning
	Interconnect Product Development
	Physical Design Tools
	Systems Design
	Engineering Change Management
	Product Development
	Tool Development
	Development Process Control
	Electronic Development
	Phase 0/Requirements

Distribution	Receiving
	Shipping
	Storage
	Field Services/Support
	Teleprocessing and Control
	Parts Expediting
	Power Vehicles
	Salvage
	Transportation
	Production Receipts
	Disbursement
	Inventory Management
	Physical Inventory Management
Financial Accounting	Ledger Control
	Financial Control
	Payroll
	Taxes
	Transfer Pricing
	Accounts Receivable
	Accrual Accounting
	Revenue Accounting
	Accounts Payable
	Cash Control
	Employee Expense Account
	Fixed Asset Control
	Labor Distribution
	Cost Accounting
	Financial Application
	Fixed Assets/Appropriation
	Intercompany Accounting/Billing
	Inventory Control
	Procurement Support
Financial Planning	Appropriation Control
	Budget Control
	Cost Estimating
	Financial Planning
	Transfer Pricing
	Inventory Control
	Business Planning
	Contract Management
	Financial Outlook

Information Systems	Applications Development Methodology
	Systems Management Controls
	Service Level Assessment
Production Control	Consignment Process
	Customer Order Services Management
	Early Manufacturing Involvement and Product Release
	Engineering Change (EC) Implementation
	Field Parts Support
	Parts Planning and Ordering
	Planning and Scheduling Management
	Plant Business Volumes Performance Management
	Site Sensitive Parts
	Systems Work in Process (WIP) Management
	Allocation
	Inventory Projection
	New Product Planning
	WIP Accuracy
	Base Plan Commit
	Manufacturing Process Record
Purchasing	Alteration/Cancellation
	Expediting
	Invoice/Payment
	Supplier Selection
	Cost
	Delivery
	Quality
	Supplier Relations
	Contracts
	Laboratory Procurement
	Nonproduction Orders
	Production Orders
	Supplier Payment
	Process Interplant Transfer

Personnel	Benefits
	Compensation
	Employee Relations
	Employment
	Equal Opportunity
	Executive Resources
	Management Development
	Medical
	Personnel Research
	Personnel Services
	Placement
	Records
	Suggestions
	Management Development/Research
	Personnel Programs
	Personnel Assessment
	Resource Management
Programming	Distributed Systems Products
	Programming Center
	Software Development
	Software Engineering
	Software Manufacturing Products
Quality	New Product Qualification
	Supplier Quality
Site Services	Facilities Change Request
Miscellaneous	Cost of Box Manufacturing Quality
	Service Cost Estimating
	Site Planning

Appendix II

Typical Output Measurements

I. Accounting Quality Measurements

1. Percent of late reports
2. Percent of errors in reports
3. Errors in input to Information Services
4. Errors reported by outside auditors
5. Percent of input errors detected
6. Number of complaints by users
7. Number of hours per week correcting or changing documents
8. Number of complaints about inefficiencies or excessive paper
9. Amount of time spent appraising/correcting input errors
10. Payroll processing time
11. Percent of errors in payroll
12. Length of time to prepare and send a bill
13. Length of time billed and not received
14. Number of final accounting jobs rerun
15. Number of equipment sales miscoded
16. Amount of intracompany accounting bill-back activity
17. Time spent correcting erroneous inputs
18. Number of open items
19. Percent of deviations from cash plan
20. Percent discrepancy in Material Review Board (MRB) and line scrap reports
21. Travel expense accounts processed in three days
22. Percent of advances outstanding
23. Percent data entry errors in accounts payable and general ledger
24. Credit turnaround time
25. Machine billing turnaround time
26. Percent of shipments requiring more than one attempt to invoice
27. Number of untimely supplier invoices processed
28. Average number of days from receipt to processing

II. Clerical Quality Measurements

1. Misfiles per week
2. Paper mailed/paper used
3. Errors per typed page
4. Administration errors (not using the right procedure)
5. Number of times manager is late to meetings
6. Number of times messages are not delivered
7. Percent of action items not done on schedule
8. Percent of inputs not received on schedule

9. Percent of coding errors on time cards
10. Period reports not completed on schedule
11. Percent of phone calls answered within two rings
12. Percent of phone calls dialed correctly
13. Pages processed error-free per hour
14. Clerical personnel/personnel supported
15. Percent of pages retyped
16. Percent of impressions reprinted

III. Product/Development Engineering Quality Measurements

1. Percent of drafting errors per print
2. Percent of prints released on schedule
3. Percent of errors in cost estimates
4. Number of times a print is changed
5. Number of off-specs approved
6. Simulation accuracy
7. Accuracy of advance materials list
8. Cost of input errors to the computer
9. How well product meets customer expectations
10. Field performance of product
11. Percent of error-free designs
12. Percent of errors found during design review
13. Percent of repeat problems corrected
14. Time to correct a problem
15. Time required to make an engineering change
16. Cost of engineering changes per month
17. Percent of reports with errors in them
18. Data recording errors per month
19. Percent of evaluations that meet engineering objectives
20. Percent of special quotations that are successful
21. Percent of test plans that are changed (change/test plan)
22. Percent of meetings starting on schedule
23. Spare parts' cost after warranty
24. Number of meetings held per quarter where quality and defect prevention were the main subject
25. Person-months per released print
26. Percent of total problems found by diagnostics as released
27. Customer cost per life of output delivered
28. Number of problems that were also encountered in previous products
29. Cycle time to correct a customer problem
30. Number of errors in publications reported from the plant and field
31. Number of products that pass independent evaluation error free
32. Number of missed shipments of prototypes
33. Number of unsuccessful preanalyses

34. Number of off-specs accepted
35. Percent of requests for engineering action open for more than two weeks
36. Number of days late to preanalysis
37. Number of restarts of evaluations and tests
38. Effectiveness of regression tests
39. Number of days for the release cycle
40. Percent of corrective action schedules missed
41. Percent of bills of material that are released in error

IV. Finance Quality Measurements

1. Percent error in budget predictions
2. Computer rerun time due to input errors
3. Computer program change cost
4. Percent of financial reports delivered on schedule
5. Number of record errors per employee
6. Percent of error-free vouchers
7. Percent of bills paid so company gets price break
8. Percent of errors in checks
9. Entry errors per week
10. Number of payroll errors per month
11. Number of errors found by outside auditors
12. Number of errors in financial reports
13. Percent of errors in travel advancement records
14. Percent of errors in expense accounts detected by auditors

V. Industrial/Plant Engineering

1. Percent of facilities on schedule
2. Percent of manufacturing time lost due to bad layouts
3. Percent of error in time estimates
4. Percent of error in purchase requests
5. Hours lost due to equipment downtime
6. Scrap and rework due to calibration errors
7. Repeat call hours for the same problem
8. Changes to layout
9. Percent deviation from budget
10. Maintenance cost/equipment cost
11. Percent variation to cost estimates
12. Number of unscheduled maintenance calls
13. Number of hours used on unscheduled maintenance
14. Number of hours used on scheduled maintenance
15. Percent of equipment maintained on schedule
16. Percent of equipment overdue for calibration
17. Accuracy of assets report
18. Percent of total floor space devoted to storage

19. Number of industrial design completions past due
20. Number of mechanical/functional errors in industrial design artwork
21. Number of errors found after construction had been accepted by the company
22. Percent of engineering action requests accepted

VI. Forecasting Quality Measurements

1. Number of upward pricing revisions per year
2. Number of project plans that meet schedule, price, and quality
3. Percent error in sales forecasts
4. Number of forecasting assumption errors
5. Number of changes in product schedules

VII. Information Systems Quality Measurements

1. Keypunch errors per day
2. Input correction on CRT
3. Reruns caused by operator error
4. Percent of reports delivered on schedule
5. Errors per thousand lines of code
6. Number of changes after the program is coded
7. Percent of time required to debug programs
8. Rework costs resulting from computer program
9. Number of cost estimates revised
10. Percent error in forecast
11. Percent error in lines of code required
12. Number of coding errors found during formal testing
13. Number of test case errors
14. Number of test case runs before success
15. Number of revisions to plan
16. Number of documentation errors
17. Number of revisions to program objectives
18. Number of errors found after formal test
19. Number of error-free programs delivered to customer
20. Number of process step errors before a correct package is ready
21. Number of revisions to checkpoint plan
22. Number of changes to customer requirements
23. Percent of programs not flow diagramed
24. Percent of customer problems not corrected per schedule
25. Percent of problems uncovered before design release
26. Percent change in customer satisfaction survey
27. Percent of defect-free artwork
28. System availability
29. Terminal response time

30. Mean time between system initial program loadings (IPLs)
31. Mean time between system repairs
32. Time before help calls are answered

VIII. Legal Quality Measurements

1. Response time on request for legal opinion
2. Time to prepare patent claims
3. Percent of cases lost

IX. Management Quality Measurements

1. Security violations per year
2. Percent variation from budget
3. Percent of target dates missed
4. Percent of personnel turnover rate
5. Percent increase in output per employee
6. Percent absenteeism
7. Percent error in planning estimates
8. Percent of output delivered on schedule
9. Percent of employees promoted to better jobs
10. Department morale index
11. Percent of meetings that start on schedule
12. Percent of employee time spent on first-time output
13. Number of job improvement ideas per employee
14. Dollars saved per employee due to new ideas and/or methods
15. Ratio of direct to indirect employees
16. Increased percent of market
17. Return on investment
18. Percent of appraisals done on schedule
19. Percent of changes to project equipment required
20. Normal appraisal distribution
21. Percent of employee output that is measured
22. Number of grievances per month
23. Number of open doors per month
24. Percent of professional employees active in professional societies
25. Percent of managers active in community activities
26. Number of security violations per month
27. Percent of time program plans are met
28. Improvement in opinion surveys
29. Percent of employees who can detect and repair their own errors
30. Percent of delinquent suggestions
31. Percent of documents that require two management signatures
32. Percent error in personnel records
33. Percent of time cards signed by managers that have errors on them
34. Percent of employees taking higher education

35. Number of damaged equipment and property reports
36. Warranty costs
37. Scrap and rework costs
38. Cost of poor quality
39. Number of employees dropping out of classes
40. Number of decisions made by higher-level management than required by procedures
41. Improvement in customer satisfaction survey
42. Volumes actual versus plan
43. Revenue actual versus plan
44. Number of formal reviews before plans are approved
45. Number of procedures with fewer than three acronyms and abbreviations
46. Percent of procedures less than 10 pages
47. Percent of employees active in improvement teams
48. Number of hours per year of career and skill development training per employee
49. Number of user complaints per month
50. Number of variances in capital spending
51. Percent revenue/expense-ratio below plan
52. Percent of executive interviews with employees
53. Percent of departments with disaster recovery plans
54. Percent of appraisals with quality as a line item that makes up more than 30 percent of the evaluation
55. Percent of employees with development plans
56. Revenue generated over strategic period
57. Number of iterations of strategic plan
58. Number of employees participating in cost-effectiveness
59. Data integrity
60. Result of peer reviews
61. Number of tasks for which actual time exceeded estimated time

X. Manufacturing and Test Engineering Quality Measurements

1. Percent of process operations where sigma limit is within engineering specification
2. Percent of tools that fail certification
3. Percent of tools that are reworked due to design errors
4. Number of process changes per operation due to errors
5. In-process yields
6. Percent error in manufacturing costs
7. Time required to solve a problem
8. Number of delays because process instructions are wrong or not available
9. Labor utilization index
10. Percent error in test equipment and tooling budget
11. Number of errors in operator training documentation

12. Percent of errors that escape the operator's detection
13. Percent of testers that fail certification
14. Percent error in yield projections
15. Percent error in output product quality
16. Asset utilization
17. Percent of designed experiments needing revision
18. Percent of changes to process specifications during process design review
19. Percent of equipment ready for production on schedule
20. Percent of meetings starting on schedule
21. Percent of drafting errors found by checkers
22. Percent of manufacturing used to screen products
23. Number of problems that the test equipment cannot detect during manufacturing cycle
24. Percent correlation between testers
25. Number of waivers to manufacturing procedures
26. Percent of tools and test equipment delivered on schedule
27. Percent of tools and test equipment on change level control
28. Percent functional test coverage of products
29. Percent projected cost reductions missed
30. Percent of action plan schedules missed
31. Equipment utilization

XI. Manufacturing/Shipping Quality Measurements

1. Complaints on shipping damage
2. Percent of parts not packed to required specifications
3. Percent of output that meets customer orders and engineering specifications
4. Scrap and rework cost
5. Suggestions per employee
6. Percent of jobs that meet cost
7. Percent of jobs that meet schedule
8. Percent of product defect-free at measurement operations
9. Percent of employees trained to do the job they are working on
10. Accidents per month
11. Performance against standards
12. Percent of utilities left improperly running at end of shift
13. Percent unplanned overtime
14. Number of security violations per month
15. Percent of time log book filled out correctly
16. Time and/or claiming errors per week
17. Time between errors at each operation
18. Errors per 100,000 solder connections
19. Labor utilization index
20. Percent of operators certified to do their job

21. Percent of shipping errors
22. Defects during warranty period
23. Replacement parts defect rates
24. Percent of products defective at final test
25. Percent of control charts maintained correctly
26. Percent of invalid test data
27. Percent of shipments below plan
28. Percent of daily reports in by 7:00 am
29. Percent of late shipments
30. Percent of products error-free at final test

XII. Marketing Quality Measurements

1. Percent of proposals submitted ahead of schedule
2. Cost of sales per total costs
3. Percent error in market forecasts
4. Percent of proposals accepted
5. Percent of quota attained
6. Response time to customer inquiries
7. Inquiries per $10,000 of advertisement
8. Number of new customers
9. Percent of repeat orders
10. Percent of time customer expectations are identified
11. Sales made per call
12. Errors in orders
13. Ratio of marketing expenses to sales
14. Number of new business opportunities identified
15. Errors per contract
16. Percent of time customer expectation changes are identified before they impact sales
17. Man-hours per $10,000 sales
18. Percent reduction in residual inventory
19. Percent of customers called back as promised
20. Percent of meetings starting on schedule
21. Percent of changed orders
22. Number of complimentary letters
23. Percent of phone numbers correctly dialed
24. Time required to turn in travel expense accounts
25. Number of revisions to market requirements statements per month
26. Percent of bids returned on schedule
27. Percent of customer letters answered in two weeks
28. Number of complaint reports received
29. Percent of complaint reports answered in three days

XIII. Personnel Quality Measurements

1. Percent of employees who leave during the first year
2. Number of days to answer suggestions
3. Number of suggestions resubmitted and approved
4. Personnel cost per employee
5. Cost per new employee
6. Turnover rate due to poor performance
7. Number of grievances per month
8. Percent of employment requests filled on schedule
9. Number of days to fill an employment request
10. Management evaluation of management education courses
11. Time to process an applicant
12. Average time a visitor spends in lobby
13. Time to get security clearance
14. Time to process insurance claims
15. Percent of employees participating in company-sponsored activities
16. Opinion survey ratings
17. Percent of complaints about salary
18. Percent of personnel problems handled by employees' managers
19. Percent of employees participating in voluntary health screening
20. Percent of offers accepted
21. Percent of retirees contacted yearly by phone
22. Percent of training classes evaluated excellent
23. Percent deviation to resource plan
24. Wait time in medical department
25. Number of days to respond to applicant
26. Percent of promotions and management changes publicized
27. Percent of error-free newsletters

XIV. Procurement/Purchasing Quality Measurements

1. Percent of discount orders by consolidating
2. Errors per purchase order
3. Number of orders received with no purchase order
4. Routing and rate errors per shipment
5. Percent of supplies delivered on schedule
6. Percent decrease in parts costs
7. Expeditors per direct employees
8. Number of items on the hot list
9. Percent of suppliers with 100 percent lot acceptance for one year
10. Stock costs
11. Labor hours per $10,000 purchases
12. Purchase order cycle time
13. Number of times per year line is stopped due to lack of supplier parts

14. Supplier parts scrapped due to engineering changes
15. Percent of parts with two or more suppliers
16. Average time to fill emergency orders
17. Average time to replace rejected lots with good parts
18. Parts cost per total costs
19. Percent of lots received on line late
20. Actual purchased materials cost per budgeted cost
21. Time to answer customer complaints
22. Percent of phone calls dialed correctly
23. Percent of purchase orders returned due to errors or incomplete description
24. Percent of defect-free supplier model parts
25. Percent projected cost reductions missed
26. Time required to process equipment purchase orders
27. Cost of rush shipments
28. Number of items billed but not received

XV. Production Control Quality Measurements

1. Percent of late deliveries
2. Percent of errors in stocking
3. Number of items exceeding shelf life
4. Percent of manufacturing jobs completed on schedule
5. Time required to incorporate engineering changes
6. Percent of errors in purchase requisitions
7. Percent of products that meet customer orders
8. Inventory turnover rate
9. Time that line is down due to assembly shortage
10. Percent of time parts are not in stock when ordered from common parts crib
11. Time of product in shipment
12. Cost of rush shipments
13. Spare parts availability in crib
14. Percent of errors in work-in-process records versus audit data
15. Cost of inventory spoilage
16. Number of bill-of-lading errors not caught in shipping

XVI. Quality Assurance Quality Measurements

1. Percent error in reliability projections
2. Percent of product that meets customer expectations
3. Time to answer customer complaints
4. Number of customer complaints
5. Number of errors detected during design and process reviews
6. Percent employees active in professional societies
7. Number of audits performed on schedule

8. Percent of quality assurance personnel to total personnel
9. Percent of quality inspectors to manufacturing directs
10. Percent of quality engineers to product and manufacturing engineers
11. Number of engineering changes after design review
12. Number of process changes after process qualification
13. Errors in reports
14. Time to correct a problem
15. Cost of scrap and rework that was not created at the rejected operation
16. Percent of suppliers at 100 percent lot acceptance for one year
17. Percent of lots going directly to stock
18. Percent of problems identified in the field
19. Variations between inspectors doing the same job
20. Percent of reports published on schedule
21. Number of complaints from manufacturing management
22. Percent of field returns correctly analyzed
23. Time to identify and solve problems
24. Percent of laboratory services not completed on schedule
25. Percent of improvement in early detection of major design errors
26. Percent of errors in defect records
27. Number of reject orders not dispositioned in five days
28. Number of customer calls to report errors
29. Level of customer surveys
30. Number of committed supplier plans in place
31. Percent of correlated test results with suppliers
32. Receiving inspection cycle time
33. Number of requests for corrective action being processed
34. Time required to process a request for corrective action
35. Number of off-specs approved
36. Percent of part numbers going directly to stock
37. Number of manufacturing interruptions caused by supplier parts
38. Percent error in predicting customer performance
39. Percent product cost related to appraisal, scrap, and rework
40. Percent skip lot inspection
41. Percent of qualified suppliers
42. Number of problems identified in-process

XVII. Security/Safety Quality Measurements

1. Percent of clearance errors
2. Time to get clearance
3. Percent of security violations
4. Percent of documents classified incorrectly
5. Security violations per audit
6. Percent of audits conducted on schedule
7. Percent of safety equipment checked per schedule

8. Number of safety problems identified by management versus total safety problems identified
9. Safety accidents per 100,000 hours worked
10. Safety violations by department
11. Number of safety suggestions
12. Percent of sensitive parts located

Bibliography

Akers, John F. Remarks from American Electronic Association on Quality. Boston, Mass., March 13, 1984.

Doran, P. K. "A Total Quality Improvement Program." Presented at 24th International Quality Assurance Conference. Oxford, England, September 1985. Published in British Quality Assurance Magazine, December 1985.

Jackson, John. "Emerging Quality Emphasis in the U.S." Presented at INSPEX/Asia 1983. Singapore, October 1983.

Kane, Ed. "IBM's Excellence Plus Program." Presented to the India Quality and Reliability Association. Bombay, India, 1984.

Kane, Ed. "IBM's Total Quality Improvement System." Excerpt from "The Quality/Profit Connection," by H. James Harrington. Milwaukee: ASQC Quality Press, 1988.

Index

Process improvement:
  teams, 32, 34, 35
  tools for, 35
Process ownership, 22, 31
Process qualification, 22, 24-25
  levels of, 23

Qualification lots, 28
Quality circles, 51
Quality councils, 12
Quality improvement:
  plans, 11
  teams, 46
Quality policy, 10

Rutgers, Samuel, 1

Strategy, 5
Structured analysis and structured design (SA/SD), 37
Structured Systems Analysis (Gane and Sarson), 37
Suppliers:
    (*see also* Business associates)
  relations with, 8, 42-44
System improvement, examples of, 41
Systems control, 7, 40-41

Tactics, 5
Targets, 20
Theory Z (Ouchi), 45
Tietze, Armin R., 37
Total participation, 8, 45

Value control (VC), 37

Watson, Thomas J. Jr., 2
Watson, Thomas J. Sr., 52